Introduction to Image Acquisition and Display Technologies

In this valuable reference work, Ichiro Fujieda focuses on the component technologies, device configurations, and operation principles of image acquisition and display technologies and provides detailed use cases to give practical guidance on the various current and potential future applications of these technologies.

The technology and the physics behind these devices can be grouped into three categories: optical technology, material science, and semiconductor device technology. This book enables readers to gain an understanding of these three areas in relation to the flow of image information and several example applications of the technology. Fujieda first describes the building blocks of image sensors and displays (detectors, light sources, transistors, and wavefront control devices) and their configurations, operation principles, and characteristics. He then describes in more detail image sensor technology (including MOS image sensors, CCD technologies, and X-ray and infrared imagers) and displays (including thin-film transistor arrays, LCDs, OLEDs, MEMS devices, and more). Finally, he provides real-world examples of how these technologies are used together to give the reader an understanding of their practical applications and their potential use in future devices. Some important laws in optics and definitions in color science are included for easy reference. Through this approach, the reader will gain a detailed understanding of each of the component parts of existing imaging devices and will be able to apply this to future developments within the field.

This book will benefit any advanced undergraduate and graduate student and industry professional who wishes to expand his or her understanding of the hardware handling digital images. Some basic knowledge is required on semiconductor device physics and the interaction of radiation with matter, though these are described in the appropriate sections.

Introduction to Image Acquisition and Display Technologies

Photon Manipulation in Image Sensors and Displays

Ichiro Fujieda

CRC Press is an imprint of the
Taylor & Francis Group, an **informa** business

Designed cover image: Ichiro Fujieda

First edition published 2024
by CRC Press
6000 Broken Sound Parkway NW, Suite 300, Boca Raton, FL 33487-2742

and by CRC Press
4 Park Square, Milton Park, Abingdon, Oxon, OX14 4RN

CRC Press is an imprint of Taylor & Francis Group, LLC

ISBN: 978-1-032-42931-1 (hbk)
ISBN: 978-1-032-42934-2 (pbk)
ISBN: 978-1-003-36497-9 (ebk)

DOI: 10.1201/9781003364979

Typeset in Bembo
by codeMantra

Contents

Preface

A beautiful picture takes our breath away. Visualizing the invisible leads to discoveries and propels our understanding of nature. There is no need to reiterate the importance of electronic systems that handle images. Image sensors and displays are essential parts of human interfaces, contributing to the evolution of our social life of information. These devices are based on electronic and photonic materials, which also provide solid foundations for radiation detection, energy harvesting, and other important systems. This book focuses on such electronic and photonic devices and systems.

Device configurations and operation principles are fundamental concepts. They remain the same until a paradigm shift occurs. New materials continue to enhance the performance of existing devices. Even if a device is once regarded impractical, it can revive with innovations in related fields. Metal–oxide–semiconductor (MOS) image sensor technology is a good example. Charge storage operation was demonstrated by G. Weckler in 1967. However, charge-coupled devices (CCDs) invented in 1970 dominated the market for many years. MOS image sensor technology made a strong comeback after steady progress in the integrated circuit industry materialized the concept of pixel-level amplification. Combining the MOS image sensor with amorphous Si (a-Si) technology resulted in a flat-panel X-ray image sensor. Combination with micro-electro-mechanical systems (MEMS) technology resulted in an uncooled infrared image sensor. Even if an idea might appear unpractical at first, a new material and combination with other technologies can change this view. Thus, device configurations and operation principles are the building blocks for future systems.

This book is intended for advanced undergraduate and graduate students as well as industry professionals who wish to expand their views on the hardware handling digital images. In addition to college-level mathematics, basic knowledge of physics is required. This includes semiconductor device physics and interaction of radiation with matters. They are described in appropriate sections. Some useful laws in optics and definitions in color science are included in the appendices.

Finally, I am grateful to many people who helped me grow personally and professionally, especially the late Prof. Victor Perez-Mendez at Lawrence Berkeley Laboratory, Dr. Robert A. Street at Xerox Palo Alto Research Laboratory, Fujio Okumura at NEC Corporation, Prof. Shigenori Yamaoka at Technology Research Association for Advanced Display Materials, and Prof. Yuzo Ono at Ritsumeikan University (all affiliations are at the time). I would also like to thank Andrew Stow of CRC Press for his continuous encouragement for writing this book.

Author

Ichiro Fujieda, PhD, is a professor in the Department of Electrical and Electronic Engineering at Ritsumeikan University, Shiga, Japan. He earned a BS from Waseda University in 1981 and an MS and a PhD from the University of California, Berkeley, in 1984 and 1990, respectively. He is a member of Optica (formerly OSA) and SPIE, the international society for optics and photonics.

1 Introduction

Various electronic and photonic devices are utilized for acquiring and displaying images. These devices and the physics behind them constitute the backbones of the systems handling digital images. By following the flow of information, three layers are identified: optical technology, materials science, and semiconductor device technology. For example, let us consider the steps involved for a digital camera to capture an image of a person. Visible light reflected by the person enters the camera. A lens system forms the image of the person on an image sensor (optical technology). The semiconductor in the pixel converts the light to electric charges (materials science). The charges generated in the pixels are read out by the transistor circuit in the image sensor (semiconductor device). This flow of information is reversed in a display. For example, the digital information is delivered to each pixel in a liquid crystal display (LCD) by transistor circuits. Liquid crystal molecules in each pixel are reoriented, and their optical characteristics change. Combined with color filters and polarizers, transmission of the light from a backlight unit is modulated.

Some technologies are applied for both acquiring and displaying images. For example, thin-film transistors (TFTs) are used to address each pixel in an LCD as well as in a flat-panel X-ray imager. Some uncooled infrared detectors in thermal cameras and digital micromirror devices (DMDs) in projectors are based on micro-electrical-mechanical system (MEMS) technology. Luminescent materials are utilized by both image sensors and displays. For example, those converting gamma rays and X-rays to visible photons are called scintillators in radiation detection. Phosphors convert blue light to yellow light in a white light-emitting diode (LED). They convert ultraviolet light to visible light in a plasma display panel (PDP). Phosphors in a cathode-ray tube (CRT) absorb electrons and emit visible light. Thus, TFTs, MEMS devices, liquid crystal and luminescent materials are examples of building blocks of image sensors and displays. By learning them, one can understand a broad range of input/output devices developed so far. Furthermore, one might invent a novel application by combining these devices and materials in a unique manner.

DOI: 10.1201/9781003364979-1

In this chapter, we look at the flow of image information and identify the building blocks for image input/output systems.

1.1 Flow of image information

Visible light carries image information in many cases. For example, images of sceneries and people are captured by digital cameras. Documents are scanned by flat-bed scanners and facsimiles. Once these images are in computers, image processing algorithms can extract some useful information. Machine learning techniques enhance recognition of characters, people, and even their facial expressions. The processed images and the information derived from them are displayed for people to appreciate (Figure 1.1).

Visible light is not the only means of carrying image information. Infrared light and X-rays are also electromagnetic radiation but with different photon energies. Other physical phenomena such as heat transfer and electrostatic capacitive coupling are also utilized for acquiring fingerprint images. Because the nature of interactions with matters is different from those for visible light, a variety of physical phenomena are utilized for detecting them. Nevertheless, making the invisible visible expands our understanding of the world and improves our life (Figure 1.2).

1.2 Three technology layers

Some technologies are commonly adopted for both acquiring and displaying images. By identifying and categorizing key technologies, one can learn the device physics in an efficient manner. Although one might be tempted to add another technology layer for image processing, pattern recognition, machine

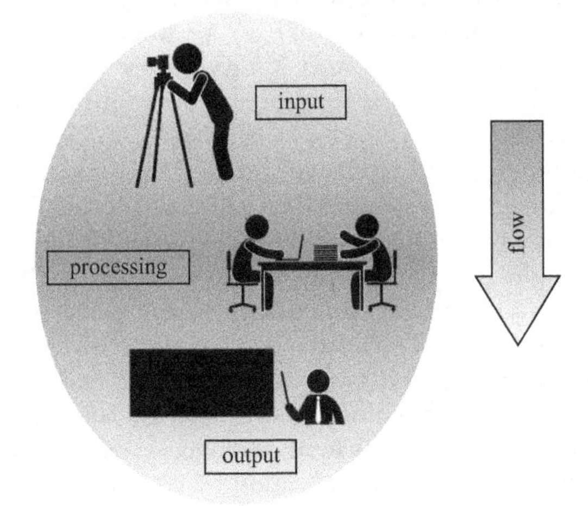

Figure 1.1 Information flow of digital images.

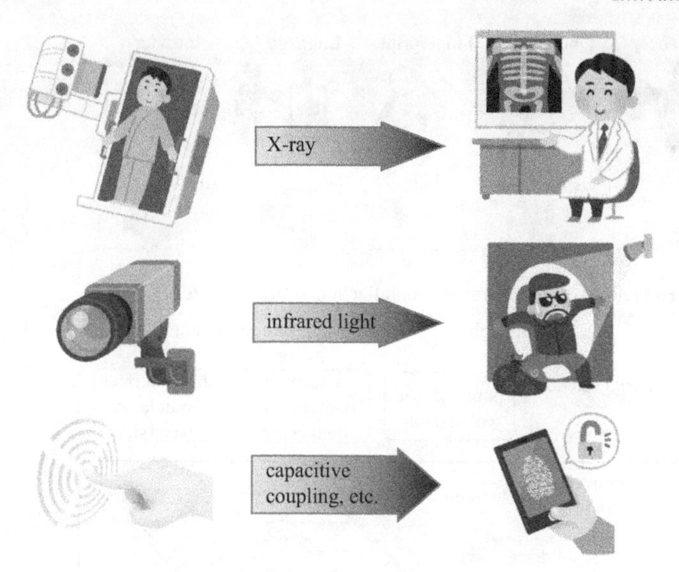

Figure 1.2 Visualizing the invisible.

learning, etc., this is outside the scope of this book. We focus on the physical aspects of image sensors and displays.

1.2.1 Image sensors

There are three technology layers for acquiring digital images as shown in Figure 1.3. Optical technologies constitute the first layer. They transfer the image information to some materials. For example, a lens relays the light from an object to the semiconductor in a photodetector. Lenses and mirrors rely on refraction and reflection of light. They are also used in microscopes and telescopes. A diffraction grating divides white light into each wavelength component. Acquiring an image for each wavelength is called hyperspectral imaging. A collimator is used to allow only the parallel X-rays transmitting an object in radiography.

In the second layer, various materials generate electric charges according to the strength of input signal. For example, high-energy photons such as X-rays and gamma rays interact with a luminescent material called scintillator in a variety of ways. In the end, visible photons are emitted. A semiconductor absorbs a visible photon and generates an electron–hole pair via photoelectric effect. An infrared photon can be converted to electric charges in the same manner by a semiconductor with a smaller bandgap energy. Alternatively, absorption of infrared photons raises the temperature of a material. Its resistance varies accordingly. By applying a bias, this information is read out as a form of electric current.

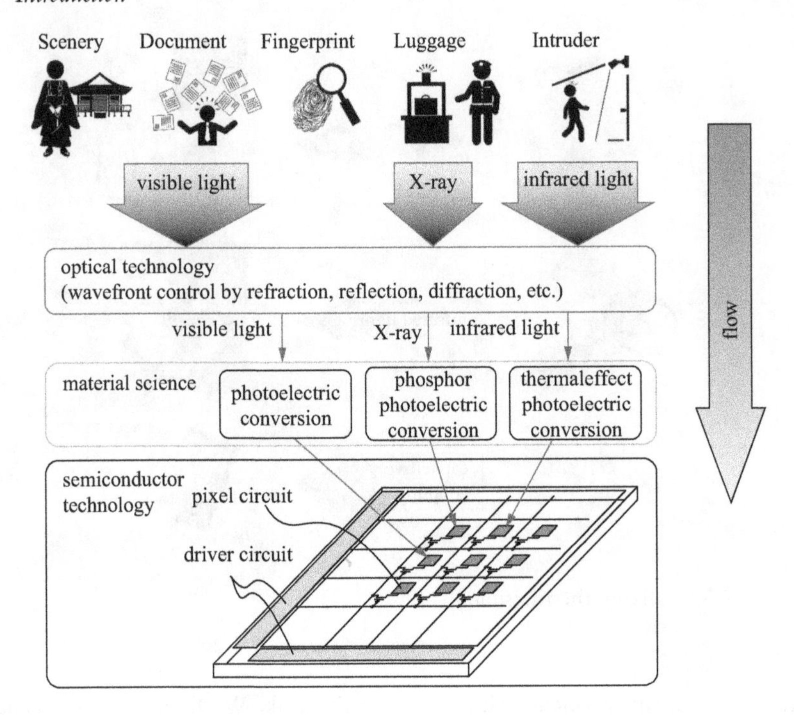

Figure 1.3 Technology layers for digital image acquisition.

One needs to transfer the electric charges stored at each pixel to an external circuit to convert them to a digital signal. Metal oxide semiconductor (MOS) technology plays an important role here, and it constitutes the third layer. For a small-area application, a metal-oxide-semiconductor field effect transistor (MOSFET) based on crystalline silicon is almost exclusively used. There are two technologies: MOS image sensors and charge-coupled devices (CCDs). Equipment for fabricating integrated circuits (ICs) is used to produce these imagers. Data processing algorithms can be incorporated into an image sensor by integrating logic ICs on the same substrate. For a large-area application such as an X-ray imager, crystalline Si wafers are too small and expensive. Hence, MOSFETs based on thin-film semiconductors are adopted. Although a whole set of different manufacturing technologies is required, concurrent development of solar cells and LCDs has benefited the field of X-ray imagers.

Nevertheless, this technology grouping is conceptional and does not necessarily represent the physical location of a material/component. For example, a photodiode is commonly used for photoconversion. A color filter is an optical device for color sensing. Both are fabricated at each pixel on an image sensor for visible light.

1.2.2 Displays

Let us examine the technologies required for displaying images. There are many display technologies, and competition among industries is harsh. For example, a CRT, a field-emission display (FED), and a PDP are no longer considered mainstream. Hence, we consider LCDs, organic light-emitting devices (OLEDs), and mini-LED displays for direct-view display applications.

As shown in Figure 1.4, there are three layers for displaying digital images: semiconductor technology, materials science, and optical technology. A similar structure is found for the case of capturing images as described above. Note that the direction of information flow is reversed.

First, digital information stored in a computer is transferred to a two-dimensional surface. For a direct-view display, it is the surface on which a final image is displayed. For a projector, this is the surface of a display device. In both cases, visible light is modulated based on this digital information. Transistor circuits are required to deliver a datum to each pixel. Thus, the semiconductor technology constitutes the first layer.

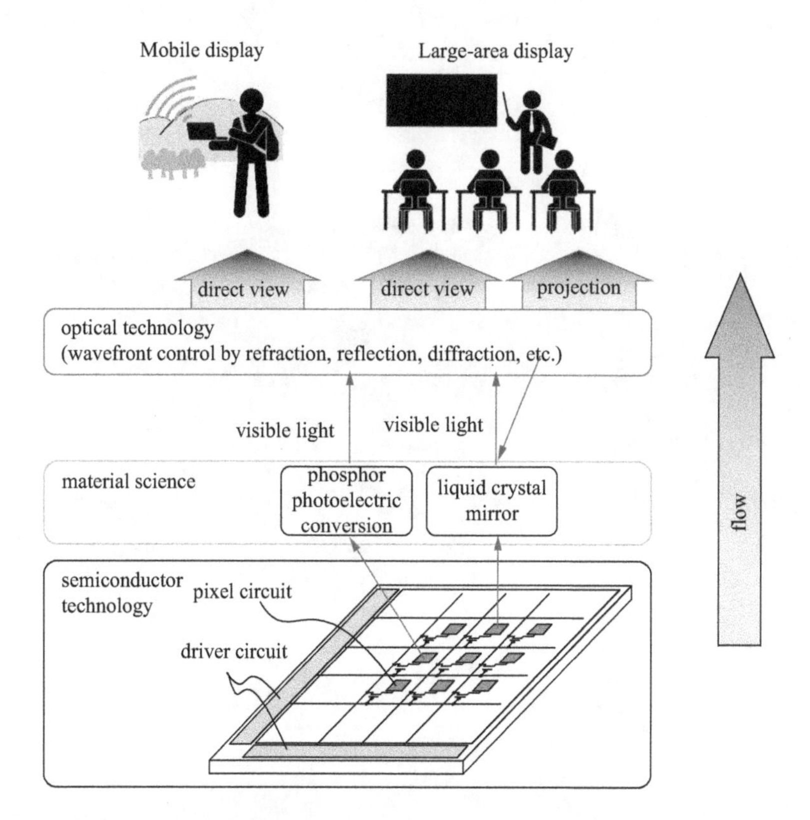

Figure 1.4 Technology layers for displaying digital images.

The second and third layers are material science and optical technology, respectively. For example, stacked organic thin films convert electric current to visible photons at each pixel in an OLED display. This is a reversed process of photoelectric effect and is called electroluminescence. The objective of liquid crystal molecules is to modulate the light from a light source in an LCD. Polarizer films are required for this transmittance control. Additive color mixing is realized by incorporating color filters and/or color-conversion materials. Incident light is modulated by a reflector at each pixel in a reflective LCD as well as in a DMD. Thus, various materials and optical devices are utilized for displaying images.

2 Electronic and photonic devices

Before inventing a novel application, one must know its building blocks. They can be categorized into four groups. First, detectors are required for acquiring images. For example, a photodiode converts a visible photon to an electron–hole pair via the photoelectric effect. Second, light sources are used for illuminating an object to be imaged. They are important components for displays as well. Third, transistors are required to control current because image information is transferred by electric charges in these devices. Finally, photons are manipulated for both acquiring and displaying images. For example, several optical films are used in a backlight unit for an LCD. Various optical filters modify characteristics of light such as propagation direction, polarization state, and wavelength. These films and filters are passive devices. In contrast, an external electric field controls liquid crystal molecules in an LCD and reflectors in a digital micromirror device (DMD). Their optical characteristics change accordingly. These active devices for wavefront control play important roles in image input/output systems.

This chapter explains electronic and photonic devices in the context of building blocks for image sensors and displays. For detailed information, readers are advised to refer to classic textbooks in these fields [1–6].

2.1 Detectors

Various imaging systems have been developed for detecting information carried by visible and invisible radiations as illustrated in Figure 2.1.

Digital cameras detect visible light. In copiers and facsimiles, a built-in light source illuminates a document. The reflected light is input by a built-in scanner. Contrast of fingerprint images can be enhanced optically. The wavelength of visible light ranges from 380 nm to 780 nm. This is defined by the International Commission on Illumination (CIE), which publishes various standards on photometry, color space, etc. A summary of color science is found in Appendix A3.

X-ray computed tomography (CT) and non-destructive inspection rely on X-ray detection. Typical X-ray energy is about 60 keV –120 keV for chest radiography and about 20 keV for mammography. In positron emission

DOI: 10.1201/9781003364979-2

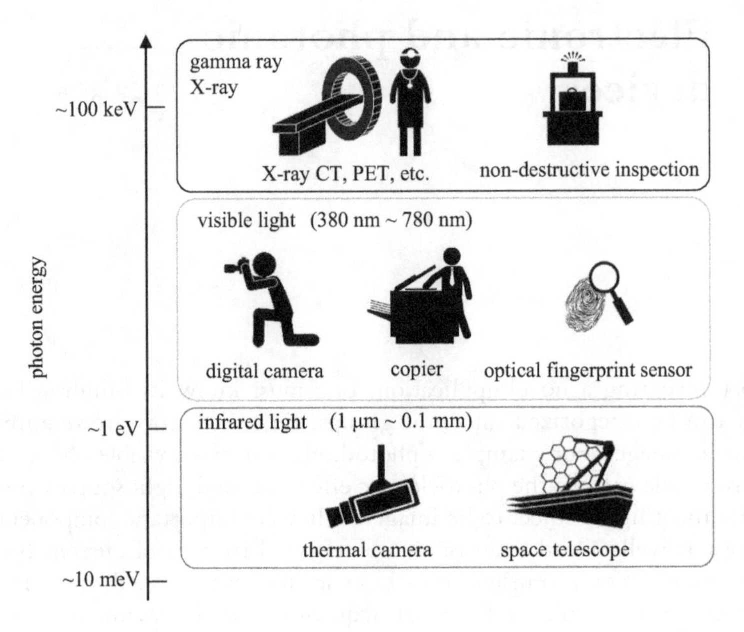

Figure 2.1 Photon energy and imaging systems.

computed tomography (PET), 511 keV gamma rays are detected to construct cross-sectional distributions of radioactive materials in a body. X-rays are emitted from an atom, while gamma rays come from a nucleus. They are both energetic photons, and their energy ranges overlap. They are indistinguishable by themselves.

Thermal cameras capture infrared images. They are used for surveillance, night vision, and military applications. Astronomers observe infrared images to understand the origin of the universe.

Visible light, gamma rays, X-rays, and infrared light are all electromagnetic radiation, namely, a photon. Quantum mechanics teaches us wave-particle duality: every entity (photons, electrons, protons, etc.) behaves like a particle and a wave. Energy E is a property of a particle. Wavelength λ is a property of a wave. Denoting the Planck constant as h and the speed of light as c, they are related by

$$E = \frac{hc}{\lambda} \tag{2.1}$$

When E is in eV and λ in nm, the product hc is equal to about 1,240. This is a useful relation to remember for converting these two quantities. Because an energetic photon behaves more like a particle, X-rays are specified by photon energy rather than wavelength. Both quantities are used for visible and infrared light.

Depending on the energy of a photon, its interaction with matter varies. Thus, detectors utilize different mechanisms and materials to convert incoming radiation to electric charges.

2.1.1 Photodiodes

A photodiode is based on the photoelectric effect. An incident photon generates an electron–hole pair in a semiconductor. A charge-sensitive amplifier connected to it converts these charges to a voltage signal. The probability of conversion (quantum efficiency) never exceeds unity. The amount of signal charges can be increased via photoconductivity as follows. Electrons generated in a material change its resistivity. When an external bias is applied, electrons can be injected. In such a case, the contact to the material is said to be Ohmic. Large current flows in the external circuit. For many applications, however, it is desired to have a signal proportional to incident photon flux. Sensors based on photoconductivity tend to have a narrower range for this linearity. Hence, photodiodes are commonly adopted for image sensors.

2.1.1.1 Configuration and operation principle

The bandgap energy of a detecting medium must be smaller than the photon energy for photoelectric effect to take place. For small-area applications such as cameras, crystalline silicon is the material of choice for detecting visible photons. Because its bandgap is about 1.1 eV, near-infrared light with wavelength shorter than about 1.1 μm can also be detected. This upper limit is called cut-off wavelength.

As illustrated in Figure 2.2, a p-n junction is formed by introducing impurity atoms to a crystalline Si wafer. The top and bottom regions are heavily doped. They are denoted as n+ and p+, respectively. They behave as metals. The top surface is usually covered by an anti-reflection coating (not shown in this figure). This structure can be fabricated as follows. A p-type layer is formed by epitaxy on a p+ substrate. Then, impurity atoms are

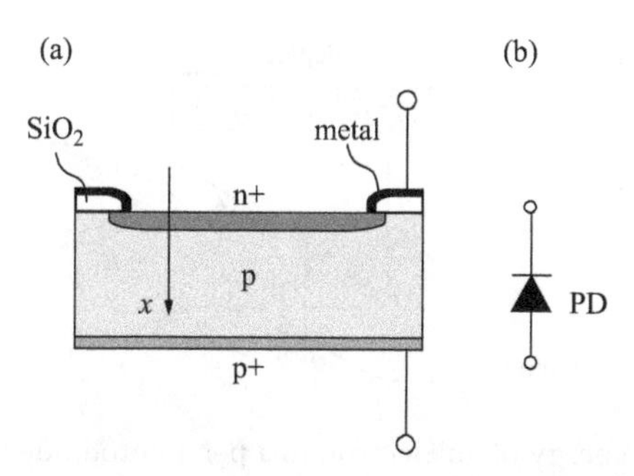

Figure 2.2 P-n junction photodiode: (a) cross-sectional view and (b) circuit symbol.

introduced by ion implantation to form an n+ layer. This is an example of a p+ Si wafer, and the doping profile can be reversed for an n+ wafer.

Mobile electrons and holes in the region around the junction are transferred such that the Fermi level becomes flat. This results in a depletion layer where impurity atoms in the Si network are ionized. They are immobile and represent fixed space charges. Due to charge neutrality, there are an equal number of positive and negative impurity atoms in the depletion layer. Hence, it extends further in a lightly doped region (the p-type layer in this example) as illustrated in Figure 2.3a. The difference in the Fermi levels of the two regions before junction formation gives rise to a built-in potential V_{bi}. Note that an electron under electric potential V has energy $E = -qV$ where q is the elementary charge. Hence, the potential difference qV_{bi} appears as shown in (a) where the conduction and valence band edges are denoted as E_C and E_V, respectively. Under the abrupt junction approximation, the potential distribution is a quadratic function of the distance along the depth direction x (see Appendix A1.1). When an external reverse bias V_{ext} is applied, the potential energy qV_{ext} is added to qV_{bi}. The depletion layer extends more in the lightly doped region as shown in Figure 2.3b.

When a photon is absorbed by the material, a pair of an electron and a hole is generated via the photoelectric effect. In the depletion region, they drift along the electric field. When they are generated in the neutral regions, they might diffuse to the depleted region and be swept away toward the electrodes.

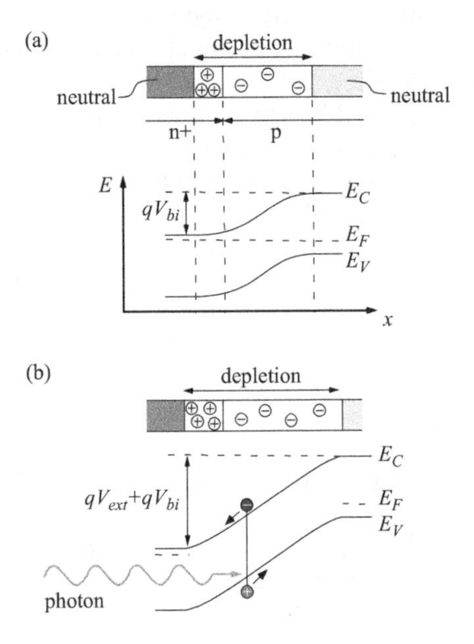

Figure 2.3 Potential energy of an electron in a p-n photodiode under (a) no external bias and (b) a reverse bias.

Instead, they might recombine. This is called geminate recombination and no signal is generated. When the reverse bias is high enough, the depletion layer extends all the way. The photodiode is under full depletion and more carriers contribute to a signal.

Electron–hole pairs are generated in a photodiode even under dark by band-to-band transitions at elevated temperatures. These charges are indistinguishable from photogenerated carriers. They result in so-called dark current. Under a large reverse bias, direct band-to-band tunneling can also cause dark current. Thermal excitation of carriers is probabilistic because dark current fluctuates with time. Thus, radiation detectors and infrared sensors based on narrow bandgap materials are susceptible to this random noise. They must be cooled to suppress this thermal excitation process.

2.1.1.2 Quantum efficiency

Quantum efficiency of a photodiode is defined as the ratio of two numbers. Denoting the number of incident photons as N_{ph} and the number of electrons at its output terminal as N_e, it is expressed as

$$\eta_{\mathrm{QE}} = \frac{N_e}{N_{ph}} \tag{2.2}$$

Quantum efficiency η_{QE} depends on the wavelength of the light. There are two mechanisms behind this. First, a photon needs to pass "window" materials. For example, ultraviolet light can be easily absorbed by a common packaging material as well as a heavily doped semiconductor. This window loss decreases η_{QE} at shorter wavelengths. Second, photon flux decreases exponentially with the distance that a photon traverses in a material. Namely, the photon flux at distance x is expressed as follows.

$$I(x) = I_0 \mathrm{e}^{-\mu x} \tag{2.3}$$

where I_0 is the flux at $x = 0$ and μ is called the absorption coefficient. This is called Lambert–Beer law. Note that μ is a function of the photon energy. For example, red light penetrates crystalline Si deeper than green light. Therefore, transmission loss causes η_{QE} to decrease at longer wavelengths.

Examples of some commercial crystalline Si photodiodes are shown in Figure 2.4. Model (S1227-BQ) has a quartz window which allows ultraviolet light to pass through. Model S5106 is a p-i-n photodiode, and its quantum efficiency remains high at longer wavelengths. Model (S1087Q) is equipped with a filter to make its curve resemble the spectral sensitivity of human eyes (luminous efficiency function). This is convenient for photometric measurement (see Appendix A3).

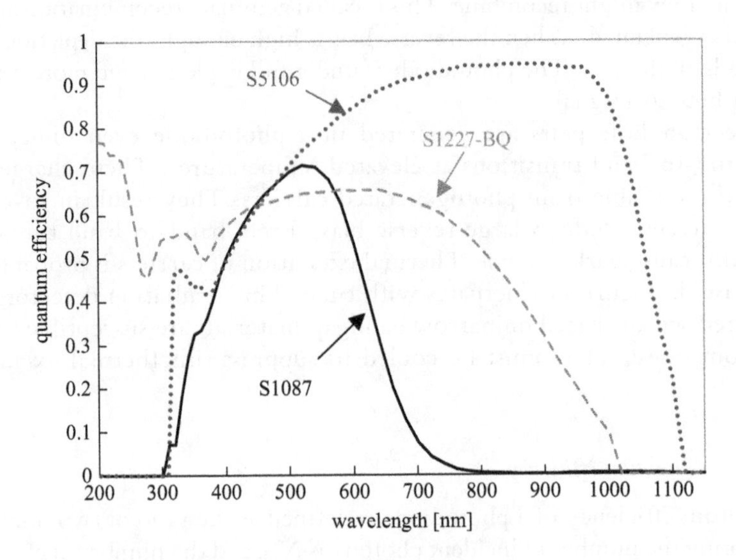

Figure 2.4 Quantum efficiency of some photodiodes. (Courtesy of Hamamatsu Photonics K.K.)

Example 2.1

When monochromatic light at 620 nm is incident on a photodiode, it records a photocurrent of 3.2 pA. The quantum efficiency of this photodiode at this wavelength is 0.50. Calculate the incident photon flux.

Solution

Photocurrent I_{ph} is equal to the number of electrons per unit time multiplied by elementary charge q. Hence, the number of electrons per unit time is equal to I_{ph}/q. Photon flux F_{ph} is equal to the number of photons per unit time. Hence,

$$\eta_{QE} = \frac{I_{ph}}{q} \cdot \frac{1}{F_{ph}} \therefore F_{ph} = \frac{I_{ph}}{q\eta_{QE}}$$

Plugging the numbers in this equation, $F_{ph} = \dfrac{3.2 \times 10^{-12}\,\text{C/s}}{1.6 \times 10^{-19}\,\text{C} \times 0.50} = 4.0 \times 10^{7}$ photons/s.

Manufacturers of photodiodes often specify responsivity R in place of quantum efficiency. This is defined as the ratio of photocurrent and incident optical power. Hence, its unit is A/W. Photocurrent I_{ph} is equal to elementary charge q multiplied by the number of electrons flowing in a unit time. Incident optical power P_{in} is equal to the energy of each photon multiplied by the number of photons incident in a unit time. Noting that each photon with wavelength λ has energy hc/λ, responsivity R is related to quantum efficiency η_{QE} as follows.

$$R = \frac{I_{ph}}{P_{in}} = \frac{qN_e}{\frac{hc}{\lambda}N_{ph}} = \frac{q\lambda}{hc}\eta_{QE} \tag{2.4}$$

2.1.1.3 *Amorphous silicon photodiode*

For large-area applications, hydrogenated amorphous silicon (a-Si:H) is adopted for the detecting medium. This is a thin-film semiconductor with a bandgap energy of about 1.8 eV. Hence, it is well suited for converting visible light to electric charges. In fact, this material had been extensively investigated for TFTs and solar cells [7,8] (see Sections 2.3.2 and 3.1). The carrier transport in a doped a-Si:H layer is much slower than in an intrinsic material. Therefore, a p-i-n configuration illustrated in Figure 2.5 is adopted for a-Si photodiodes and solar cells. A transparent top electrode and a bottom metal electrode ensure uniform potential distribution across a large sensitive area.

Moving charges induce charges with opposite signs at each electrode. This is known as the Ramo theorem [9]. This photocurrent flows while they are moving. The drift speed is proportional to the electric field $\mathcal{E}$, and the proportionality constant is called mobility. Using subscripts, drift mobility for each carrier is expressed as follows.

$$\begin{aligned} v_e &= \mu_e \mathcal{E} \\ v_h &= \mu_h \mathcal{E} \end{aligned} \tag{2.5}$$

Mobilities of electrons and holes in a-Si:H films are evaluated by time-of-flight (TOF) experiments [10]. A bias is applied for a short duration of time to

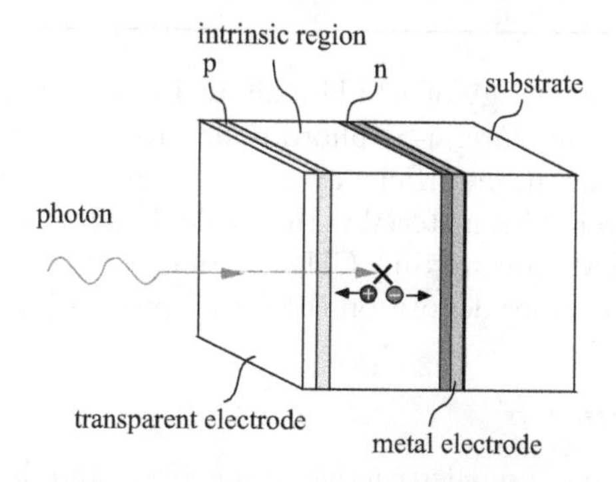

Figure 2.5 Structure of a "p-i-n" a-Si:H photodiode.

prevent dangling bonds from trapping the carriers (see Section 3.1.2). Under this condition, there are no fixed charges in the film and the electric field is constant throughout the film. A very short pulse of light illuminates the film. Its wavelength is chosen such that the light is absorbed at the surface completely. One type of photogenerated carriers reaches one edge of the film immediately. The carriers with the opposite sign transverse the entire thickness of the film at a constant speed given by the equations above. An oscilloscope is used to record the charges induced by the moving carriers. Due to dispersion of the moving charges, the output waveform is not a rectangle. But one can identify a clear kink in the waveform to determine the transit time. Electron mobility in a-Si:H films is about three orders of magnitude lower than the mobility in crystalline Si. However, it is adequate for driving large-area imagers and displays because of their operation principles to be described in the following chapters.

Example 2.2

A 1.0 μm-thick p-i-n a-Si:H photodiode is reverse-biased at 5.0 V. Let us assume that it is fully depleted and that the electric field is uniform throughout the entire device. The transit time of the electron generated at the edge of the device is 2.0 ns. Calculate the electron mobility in the intrinsic a-Si:H layer.

Solution

Denoting the electric field and the bias across the distance d as $\mathcal{E}$ and V, respectively, the transit time is expressed as, $t = \frac{d}{\mu\mathcal{E}} = \frac{d}{\mu V/d} = \frac{d^2}{\mu V}$.

Solving this for the mobility and plugging in the numbers,

$$\mu = \frac{d^2}{Vt} = \frac{\left(1.0 \times 10^{-4}\,\text{cm}\right)^2}{5.0\text{V} \times 2.0 \times 10^{-9}\,\text{s}} = 1.0\ \text{cm}^2/\text{V} \cdot \text{s}$$

Because the bandgap energy of a-Si:H is about 1.8 eV, the cut-off wavelength is about 690 nm. Therefore, a-Si photodiodes are not sensitive to infrared light. To enhance quantum efficiency at shorter wavelengths, a-SiC can be used for the p+ layer. This material with a wider bandgap energy is deposited on the intrinsic layer by mixing CH_4 to SiH_4 and B_2H_6 in the plasma-enhanced chemical vapor deposition (PECVD) process [11].

2.1.2 Radiation detectors

In medical imaging, non-destructive inspection, and high-energy physics experiments, X-rays and gamma rays are detected. They possess much higher energy than visible light. There are two strategies for detecting them: direct and indirect methods. In the direct detection scheme, photoelectric effect in a semiconductor is utilized. Electron–hole pairs are generated and their movement in a diode structure generates output signals as in the case of photodiodes. Alternatively, one can convert X-rays and gamma rays to visible photons with luminescent materials and use photodiodes to detect them. Such luminescent materials are called scintillators in the field of radiation detection.

Before describing semiconductor detectors and scintillation detectors, a review of some physics and electronics related to radiation detection is in order.

2.1.2.1 Interaction of radiation with matter

Three types of interactions are relevant: photoelectric effect, Compton scattering, and pair production. These are illustrated in Figure 2.6. First, an energetic photon ejects an electron from an atom via photoelectric effect as shown in (a). The ejected electron (photoelectron) is energetic and induces subsequent interactions. Second, an energetic photon collides with the electron in an atom. As shown in (b), a scattered photon and an electron emerge from the atom. This is called Compton scattering. Energy and momentum are conserved before and after the collision. Note that the energy of the scattered X-ray is lower than that of the incident X-ray. This feature can be utilized to reject scattered X-rays in imaging applications. Third, when a photon passes by a nucleus, its energy can be converted to a pair of an electron and a positron as shown in (c). This is called pair production. An electron and a positron have the same mass. Its rest-mass energy is $m_e c^2 = 511$ keV. The law of energy conservation requires that the incoming photon must possess energy larger than $2m_e c^2 = 1.02$ MeV. The nucleus is recoiled to conserve momentum.

After these initial interactions, subsequent events occur as illustrated in Figure 2.7. After photoelectric effect, an electron in outer shells fills the vacant state. The excess energy is released as a photon. Its energy is specific to the atom and the photon is called characteristic X-ray. When an electron passes

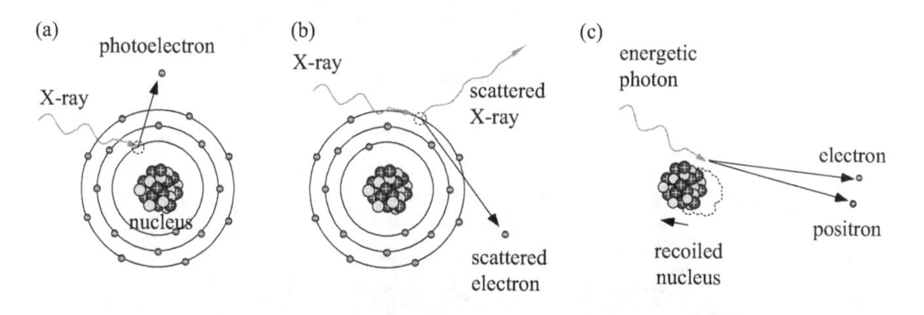

Figure 2.6 Interaction of an energetic photon with matter: (a) photoelectric effect, (b) Compton scattering, and (c) pair production.

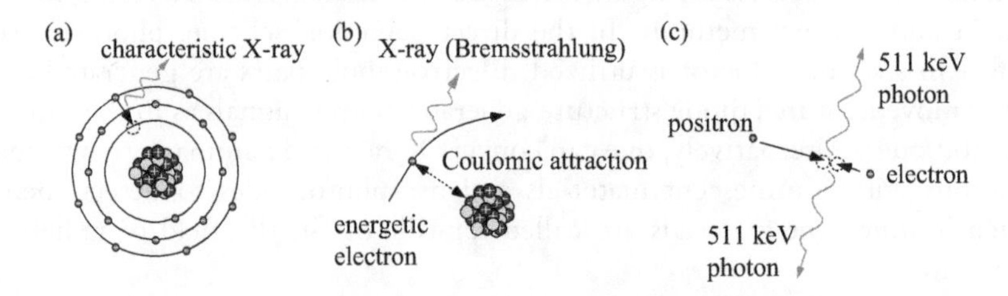

Figure 2.7 Secondary events leading to emissions of high-energy photons: (a) characteristic X-ray, (b) Bremsstrahlung, and (c) pair annihilation.

around a nucleus, Coulombic attraction changes its trajectory. X-rays called Bremsstrahlung emerge in this process. Its spectrum is continuous. When a positron meets an electron, they are converted to two 511 keV photons. They are emitted in opposite directions. The angle between them deviates from 180° slightly due to momentum conservation. This phenomenon is called pair annihilation.

The resultant electrons and photons are likely to be energetic enough to cause further interactions. If the photon energy exceeds 1.02 MeV, pair production can occur as well. These interactions continue until the energy of the initial photon is dissipated in a material. As a result, many electrons are generated by a single high-energy photon. The whole process is referred to as energy cascade.

The next question is how probable these interactions are in a specific material. For quantitative analysis of these statistical events, the concept of cross section is introduced. As depicted in Figure 2.8, let us consider a disk of area σ [cm^2] distributed in space. When a particle hits this disk, we regard that the interaction occurs. Denoting the density of the disks as n [cm^{-3}], the number of disks in a small volume of area A [cm^2] and thickness Δx [cm] is expressed as $nA\Delta x$. Here, we assume that the distance Δx is small and that the disks do not overlap. Then, the total area of the disks in this slice is equal to $\sigma nA\Delta x$. Let the incident particle flux be I $\left[\text{particles/s}\right]$. Then, the rate of

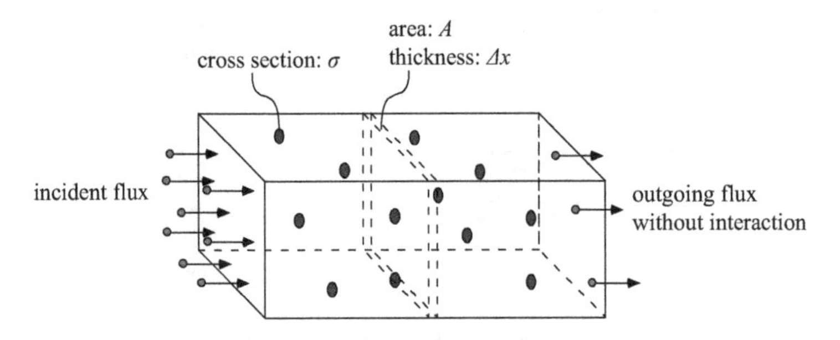

Figure 2.8 Particle flux and cross section for interaction.

interaction in this volume is $\sigma nA\Delta x \cdot I$ [s^{-1}]. This rate is for all the incident particles $A \cdot I$ [s^{-1}] and for the total number of disks $nA\Delta x$. Hence, the probability of interaction for a single particle by a single disk is given by

$$\frac{\sigma nA\Delta x \cdot I}{A \cdot I \cdot nA\Delta x} = \frac{\sigma}{A} \tag{2.6}$$

Let us consider the change in the particle flux in this infinitesimally short distance Δx [cm]. The number of disks in this slice is $nA\Delta x$. Thus, the probability of interaction for a single particle is $\frac{\sigma}{A} \cdot nA\Delta x = n\sigma\Delta x$. This is equal to the fractional change in the particle flux: $-\frac{I(x+\Delta x) - I(x)}{I(x)}$. The minus sign means loss of particles. Hence,

$$\frac{dI}{dx} = -n\sigma I \tag{2.7}$$

Introducing a new parameter by $\mu \equiv n\sigma$, solving this differential equation results in the Lambert–Beer law. The unit of μ is cm^{-1}. Because the flux is attenuated by this factor per unit length, this is called linear attenuation coefficient. Its inverse μ^{-1} represents the most likely distance that a particle moves in the material before an interaction. For this reason, it is called mean free path.

Let us apply this analysis to the case of interactions of radiation with matter. There are three kinds of interactions: photoelectric effect, Compton scattering, and pair production. Because they are independent, the total cross section for these interactions is expressed as

$$\sigma_{total} = \sigma_{pe} + \sigma_C + \sigma_{pp} \tag{2.8}$$

where each subscript represents the corresponding interaction. This is rewritten with linear attenuation coefficients as follows.

$$\mu_{total} = \sigma_{total} n \equiv \mu_{pe} + \mu_C + \mu_{pp} \tag{2.9}$$

Furthermore, it is customary to divide the coefficients by the density of a material ρ and call the resultant parameter mass attenuation coefficient.

$$\mu_m = \frac{\mu}{\rho} \tag{2.10}$$

Mass attenuation coefficients are available to the public. Data for selected materials are reproduced as a function of photon energy in Figure 2.9. At lower energy, the photoelectric effect is dominant. Compton scattering

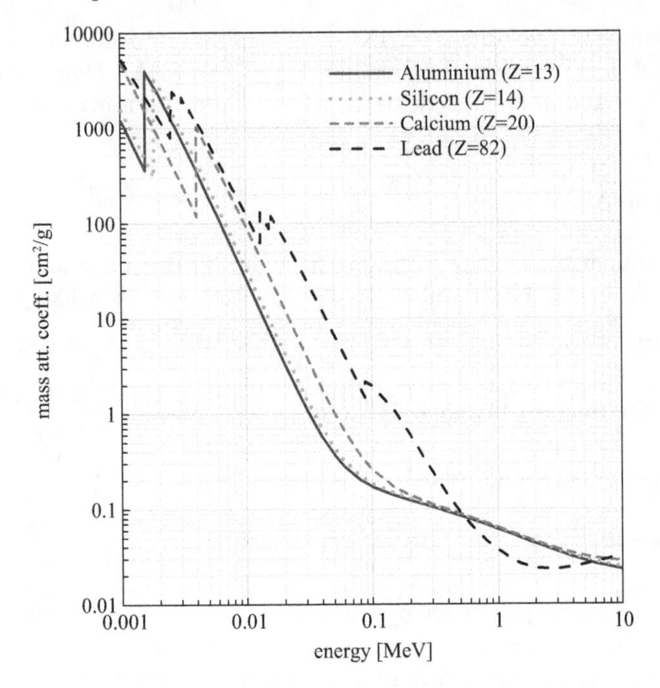

Figure 2.9 Mass attenuation coefficients for selected materials [12].

contributes more at a higher energy. Pair production starts to contribute beyond 1.02 MeV.

Note that there is a kink in the curves for Al, Si, and Ca. This sudden change is called K-edge absorption. As the photon energy reaches a level high enough to eject K-shell electrons, the probability of photoelectric effect increases abruptly. As the atomic number Z increases, the binding energy of these electrons increases. Hence, K-edge shifts to higher energy. There are multiple kinks in the curve for Pb. Its K-edge is at 88 keV. The edges at lower energies are due to photoelectric effect involving outer-shell electrons.

If cross sections of each interaction at all energies are known for a material, one can follow a particle by using its mean free path and its angular probability distribution for each interaction. Which interaction occurs is determined by generating a random number between 0 and 1 and comparing it with the probability of each interaction. This analysis is known as the Monte Carlo simulation.

2.1.2.2 Nuclear electronics

X-ray pulses are generated by an X-ray tube in practical applications. Its timing, time duration, and energy are well controlled. There are lots of X-ray photons in such a pulse. Electric charges generated by a detector are

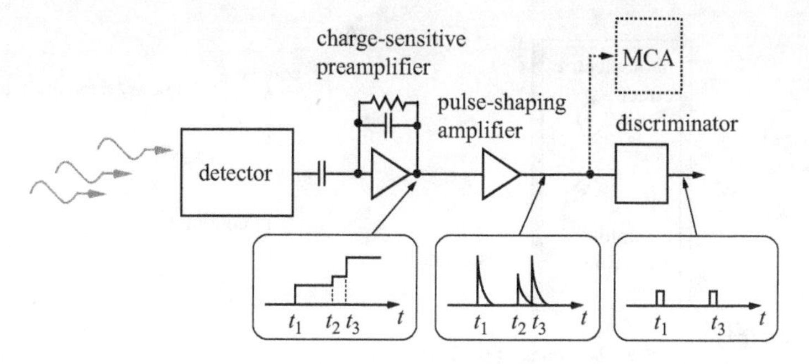

Figure 2.10 Components of nuclear electronics and waveform at each stage.

converted to a voltage by an external charge-sensitive amplifier just as in the case of a photodiode detecting visible light.

In contrast, gamma rays are emitted by a decaying nucleus. Because decay of radioisotopes is a stochastic event, gamma rays are emitted one by one. A single gamma ray is detected by nuclear electronics. Its components and waveform at each stage are illustrated in Figure 2.10. Suppose that three photons enter the detector at random intervals and deposit some of their energies. The charge-sensitive preamplifier outputs a voltage proportional to the energy deposited by each photon. This voltage adds up for each event if the intervals are much shorter than the time constant of this circuit. The pulse-shaping amplifier trims the waveform such that the pulses do not pile up. The next circuit is called discriminator. It generates a digital pulse only if the input voltage is in a certain range. This process is called energy discrimination and is used to eliminate scattered photons in imaging applications. For pulse height analysis in spectroscopy applications, an instrument called multi-channel analyzer (MCA) is used to record a histogram of the pulse heights. The resultant plot is called a pulse height spectrum. It shows the likelihood of an incident photon deposits a part of its energy.

2.1.2.3 Semiconductor detectors

Basic configuration of a semiconductor radiation detector is equivalent to a photodiode [13]. A planar configuration is illustrated in Figure 2.11a. The contacts must be a blocking type. When a reverse bias is applied, the material should be fully depleted. Charge carriers are generated throughout its detecting medium by the energy cascade process described above. A high-purity material is desired because impurities and defects can trap charge carriers. Then, the carriers can travel all the way to the corresponding electrodes, resulting in a larger signal.

When photoelectric effect occurs in the detector, it is likely that the resultant photoelectron deposits its energy nearby. Characteristic X-ray and

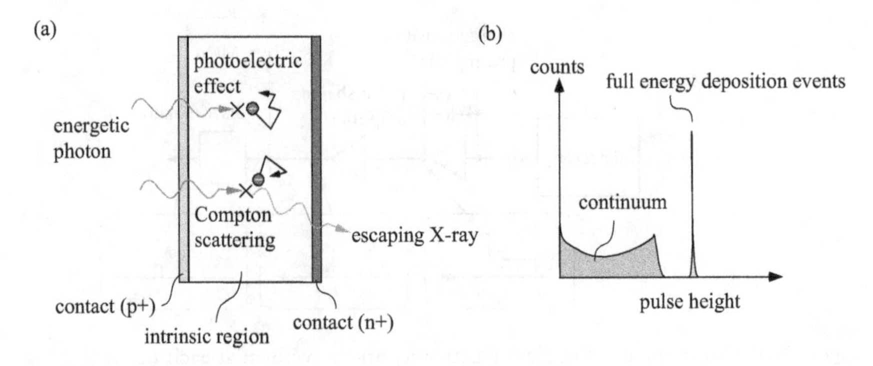

Figure 2.11 Operation principle of a semiconductor radiation detector: (a) cross-sectional view of a planar configuration and (b) simplified pulse height spectrum.

Bremsstrahlung might escape from the detector. In the case of Compton scattering, the scattered X-ray might escape, leading to partial energy deposition. As schematically shown in Figure 2.11b, its output shows a sharp peak corresponding to full-energy deposition and a continuum due to events with partial energy deposition. Relative magnitudes of these features depend on the volume of the detecting medium and the incoming radiation. Coaxial configuration is adopted by a detector based on high-purity Ge to increase its volume. Such detectors are applied for gamma-ray spectroscopy.

Various semiconductor materials are utilized for radiation detection. The average energy required for creating an electron–hole pair in a semiconductor is referred to as ε value, W value, quantum yield, radiation-ionization energy, pair creation energy, etc. This value is important in spectroscopy and nuclear medicine where the energy of an incoming photon is of interest. High energy resolution is important for identifying elements in material analysis and for acquiring clear images in nuclear medicine.

The energy of an incoming photon is dissipated for ionization and lattice vibration (phonon). The former component should be proportional to the bandgap energy of a semiconductor. The latter should be constant. Based on this consideration, Klein proposed a phenomenological model for ε values [14]. Denoting bandgap energy as E_g, this is expressed as,

$$\varepsilon = \frac{14}{5}E_g + r, \text{ where } 0.5 \le r \le 1.0 \text{ eV} \tag{2.11}$$

A review paper published in 2006 on this subject shows a different linear relation as follows [15].

$$\varepsilon = 2E_g + 1.43 \text{ eV} \tag{2.12}$$

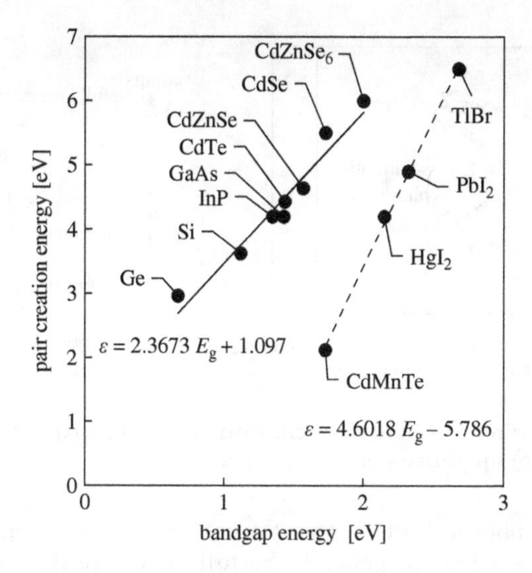

Figure 2.12 Average energy required for creating an electron–hole pair in semiconductors.

Experimental ε values in the review paper published in 2008 [16] are plotted in Figure 2.12. Interestingly, two linear relationships exist. For the materials containing atoms with high atomic numbers ($_{82}$Pb, $_{81}$Tl, $_{80}$Hg) and the compound semiconductor CdMnTe [17], the gradient of the linear fit is larger than the two phenomenological models and its intercept is negative. This indicates a potential need for more theoretical works.

The semiconductors mentioned above are all crystalline. For large-area applications such as chest X-ray imaging, tiling small discrete devices poses a problem for connecting readout electronics. In this regard, amorphous selenium (a-Se) is advantageous because it can be evaporated over a large area. Its atomic number is 34. A relatively thick (~1 mm) layer and a high voltage can ensure a decent detection efficiency for X-ray photons at ~60 keV. Its ε value is reported to depend on the electric field. For example, it is about 50 eV at the field of 10 V/μm, which is a typical operating field for a-Se flat-panel X-ray imagers [18]. This ε value is at least one order of magnitude larger than crystalline materials.

2.1.2.4 Scintillation detectors

As illustrated in Figure 2.13a, a scintillator is coupled to a photodetector in a scintillation detector. When an energetic photon excites the scintillator, it emits light in the visible wavelength range. A photodetector such as a photodiode converts these scintillation photons to a signal. Depending on the volume and the property of the scintillator, full and partial energy deposition

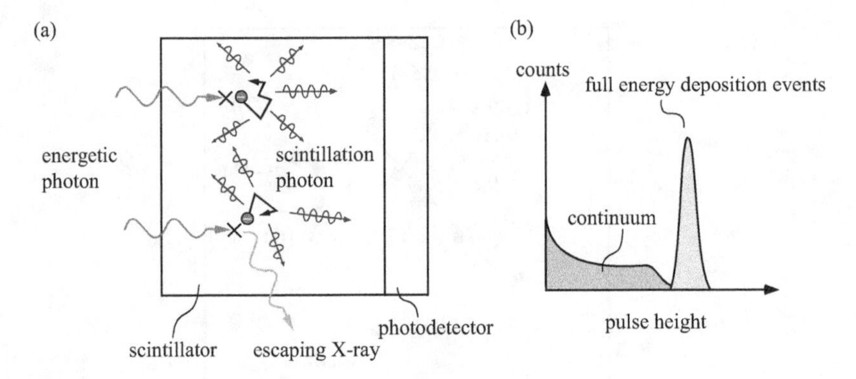

Figure 2.13 Operation principle of a scintillation detector: (a) cross-sectional view and (b) simplified pulse height spectrum.

by the incident photon leads to two features in its pulse height spectrum as shown in Figure 2.13b. In general, the full-energy peak is wider than the case for a semiconductor detector and it overlaps with the continuum due to Compton scattering. The origin of this peak widening is the statistical fluctuations in the number of scintillation photons created by a single energetic photon.

Photon yield is an important property for a scintillator because it sets an upper limit for the signal size. It is defined as the number of scintillation photons divided by the energy absorbed. Another important parameter for a scintillator is its density. Because photoelectric effect is more likely to occur in the materials with high atomic numbers, such scintillators are suited for stopping X-rays and gamma rays. Paying attention to these two properties, data from a review paper by Milbrath et al. are plotted in Figure 2.14. Most scintillators in this figure contain heavy elements. An exception is a plastic scintillator, which is adopted in high-energy physics experiments, astrophysics, national security, etc. Incorporation of heavy elements is reported to sensitize plastic scintillators [19].

Small crystalline scintillators are coupled to photodetectors in discrete radiation detectors for PET and single photon emission computed tomography (SPECT) in nuclear medicine. In the 1980s, a dense but not-so-bright scintillator bithmuth germanate, $Bi_4Ge_3O_{12}$ (BGO) was investigated for detecting 511 keV photons for PETs. A pair of discrete detectors outputting signals simultaneously determines the line on which a positron emitter exists. Extending this idea of coincidence measurement further, TOF information from the detector pair can be utilized for determining the position of the decaying nucleus on such a line. Fast scintillators such as Ce-doped lutetium oxyorthosilicate LSO(Ce) have been intensively investigated for such a TOF-PET [20].

Clinically, a dominant scintillator in nuclear medicine is still a planar Tl-doped sodium iodide NaI(Tl) crystal adopted in the Anger camera [21].

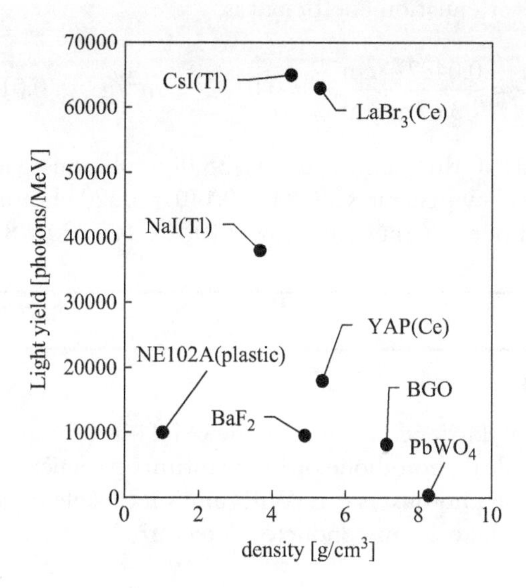

Figure 2.14 Light yield and density of scintillators.

Hal Anger invented the camera in 1952. Multiple photomultiplier tubes (PMTs) are coupled to a large planar crystal of NaI(Tl), which is needed to cover a whole human body. When scintillation photons are created by a single gamma ray, they spread in the crystal and enter the PMTs nearby. Centroid calculation of the signals from these PMTs determines the incident position of the gamma ray.

Example 2.3

A widely used radioisotope in nuclear medicine is technetium 99m $\left({}^{99\mathrm{m}}\mathrm{Tc}\right)$ [22]. It emits 140 keV gamma rays. Suppose that a 12.7 mm-thick NaI(Tl) crystal stops 94% of them. An ideal photodiode of 1.0 quantum efficiency is used to detect scintillation photons.

Calculate the mass attenuation coefficient of this scintillator at 140 keV.

How many photons are created by a single 140 keV gamma ray?

Suppose that all the scintillation photons enter the photodiode. What is the magnitude of the signal charge?

Solution

From the Lambert–Beer law, $I(x) = I_0\,\mathrm{e}^{-\mu x}$

$$\therefore \mu = -\frac{1}{x}\ln\frac{I}{I_0} = -\frac{1}{1.27\mathrm{cm}}\ln 0.94 = 0.0487\cdots\mathrm{cm}^{-1} \therefore 0.049\mathrm{cm}^{-1}$$

Hence, the mass attenuation coefficient is,

$$\mu_m = \frac{\mu}{\rho} = \frac{0.0487\cdots \text{cm}^{-1}}{3.67\,\text{g/cm}^3} = 0.0132\cdots \text{cm}^2/\text{g} \quad \therefore 0.013\,\text{cm}^2/\text{g}$$

The light yield of this scintillator is 38,000 photons/MeV. An energy deposition of 140 keV generates $38{,}000 \times 0.140 = 5{,}320$ photons in the crystal.

The photodiode output is $1.60 \times 10^{-19}\text{C} \times 5{,}320 = 8.512 \times 10^{-16}\text{C} =$ 0.85. fC

Example 2.4

The light yield of Tl-doped cesium iodide CsI(Tl) is 6.5×10^4 photons/MeV. Suppose that an ideal photodiode of 1.0 quantum efficiency is used to convert all the scintillation photons. Is this configuration capable of generating signal charges comparable to a semiconductor detector?

Solution

Because absorbed radiation energy of 1MeV creates 6.5×10^4 photons in the photodiode, ε value for this configuration is $\dfrac{1.0 \times 10^6}{6.5 \times 10^4} = 15.3\cdots$electrons/eV.

This is smaller than the ε value of amorphous selenium (a-Se) under a bias of 10 V/μm, which is about 50 eV (see Section 4.3.3). On the other hand, all the crystalline semiconductors in the graph above have smaller ε values.

2.1.2.5 Photodetectors for scintillators

A highly sensitive photodetector is required for single photon counting. An internal multiplication process is desired. For this reason, a PMT has been used to detect scintillation photons. For many applications including imaging, vacuum devices are being replaced by solid-state counterparts. Photodetectors for scintillation detectors are no exception. An avalanche photodiode (APD) multiplies signal charges via impact ionization inside the device under a large reverse bias near breakdown. Hence, its stability and uniformity are problematic. Because dark current is also multiplied, the signal-to-noise ratio is an issue. An interesting solution is provided by a device called Si photomultiplier (SiPM), solid-state photomultiplier (SSPM), multipixel photon counter (MMPC), etc. [23]. The appropriate name might be MMPC because it is essentially an array of APDs with a common load. Each "pixel" is operated at "Geiger mode," i.e., it outputs "1" if a photon is incident. The common load gives a signal proportional to the number of pixels with output "1." If the number of "pixels" is large enough, one can regard an MMPC as an analog device. It has been applied to PET, SPECT, and high-energy physics experiments [24].

Photodiodes can be used with a scintillator for X-ray detection. In this regard, the emission spectrum of CsI(Tl) matches the quantum efficiency curve of an a-Si:H photodiode well. Its light yield is higher as shown in the plot above. Furthermore, evaporated CsI(Tl) layers have columnar structures [25]. The diameter of the columns varied from about 10 μm to 20 μm by raising the substrate temperature during evaporation from 100°C to 200°C. Hence, one can increase the thickness of the layer for higher detection efficiency without sacrificing spatial resolution. Such a structured CsI(Tl) layer was also coupled to a charge-coupled device (CCD) for non-destructive inspection [26].

2.1.3 Infrared detectors

Commonly used acronyms for infrared (IR) wavelength ranges are NIR (near IR from 0.8 μm to 1.4 μm), SWIR (short-wavelength IR from 1.4 μm to 3 μm), MWIR (medium-wavelength IR from 3 μm to 8 μm), LWIR (long-wavelength IR from 8 μm to 14 μm) and VLWIR (very long-wavelength IR from 14 μm to μm 30 μm). Illustrated in Figure 2.15 are the technologies for approximate wavelength ranges. There is an absorption band in air around 5 μm–8 μm. Atmospheric transmittance deserves attention for imaging applications on earth.

There are two types of IR detectors: quantum detectors and thermal detectors. Quantum detectors are based on the photoelectric effect. For detecting NIR photons, crystalline Si, $In_xGa_{1-x}As$, and organic materials can be used. For longer wavelengths, either narrow-band semiconductors or crystalline Si with impurities are used. In these materials, cooling is required to suppress dark current caused by thermal excitation of carriers. Uncooled

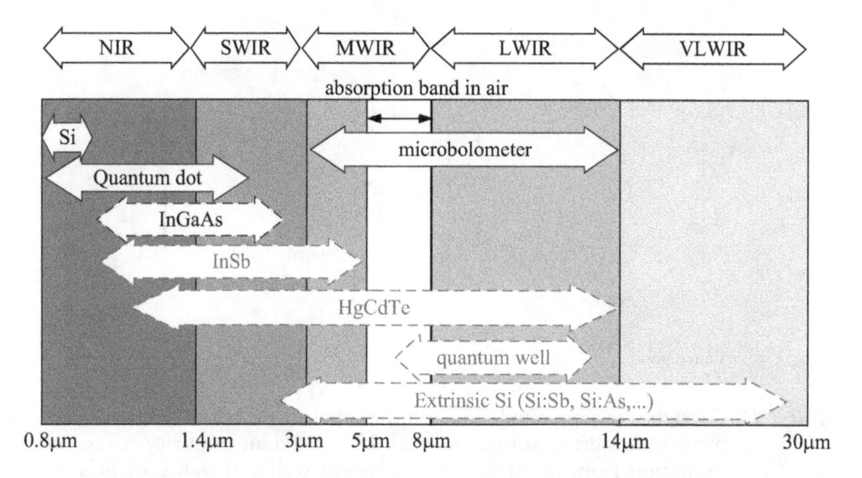

Figure 2.15 Acronyms for wavelength ranges and technologies for infrared detectors.

detectors are desired for low-cost applications. In a thermal detector, a small temperature change caused by absorption of IR photons is converted to an electrical signal. Micro-electrical-mechanical system (MEMS) technology allows one to fabricate small heat capacitors well isolated from the surrounding environment. They are called microbolometers.

2.1.3.1 Quantum detectors

As in the case of visible light, IR photons can generate carriers in a semiconductor via the photoelectric effect. For the wavelength range up to 1.1 μm, crystalline Si photodiodes can be used. An incident photon excites an electron in the valence band as schematically shown in Figure 2.16a. The electrons generated by this band-to-band transition contribute to signal generation. Compound semiconductor $In_xGa_{1-x}As$ can cover the range of 1 μm–2.6 μm. Fiber-optic communication utilizes high-speed $In_{0.53}Ga_{0.47}As$ photodetectors that operate in the 1 μm–1.7 μm wavelength range [27]. Colloidal quantum dots based on lead sulfide (PbS) have been applied to extend the wavelength range to 2 μm [28]. Absorbance of IR photons can be changed by tuning the size of quantum dots.

For longer wavelengths, a semiconductor with a narrower bandgap energy is used. The bandgap of indium antimonide (InSb) is 0.17 eV. Hence, its cut-off wavelength is 7.29 μm. Photodiodes based on this semiconductor are used for the wavelength range of 1 μm–5 μm. They must be cooled to 77 K, the boiling point of liquid nitrogen. Large InSb detector arrays were developed for missions for the James Webb Space Telescope [29]. Mercury cadmium telluride ($Hg_{1-x}Cd_xTe$) has a long history of development. Depending on its composition, its bandgap varies. Various techniques are used to form p-n junctions and these photodiodes detect IR photons with wavelength ranging from 1 μm to 10 μm and beyond [30].

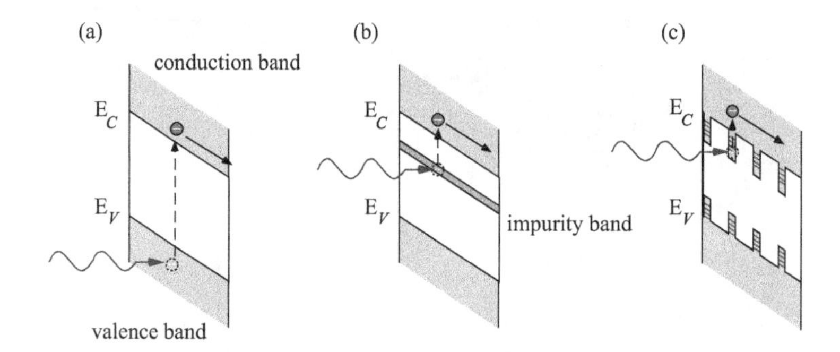

Figure 2.16 Carriers are generated in a semiconductor via photoelectric effect: (a) band-to-band transition, (b) excitation from an impurity band, and (c) excitation from the states in a quantum well and transport in a MQW structure.

For even longer wavelengths, impurity atoms such as arsenic (As) [31] and antimony (Sb) [32] are introduced in crystalline Si to form an impurity band. As depicted in Figure 2.16b, an incident photon can excite an electron in this band via the photoelectric effect into the conduction band. Detectors based on this principle are called blocked impurity band (BIB) detectors, impurity band conductor (IBC) detectors, extrinsic Si detectors, etc. A $1,024 \times 1,024$ Si:As IBC detector is installed in the Mid-Infrared Instrument (MIRI) for the James Webb Space Telescope [33].

There is another type of detector called a quantum-well infrared detector (QWIP) [34]. An IR photon absorption leads to the intra-band transition of electrons in quantum wells. As shown in Figure 2.16c, the excited electrons can escape the well in a multiple quantum well (MQW) structure. Among various materials, the technology based on GaAs/AlGaAs MQW is the most mature [35].

2.1.3.2 Thermal detectors

Absorption of radiation raises the temperature of a material. At high temperatures, electrons are scattered more by the network of constituent atoms (lattice). The drifting electrons are slowed down and the electrical resistance of the material increases. This change can be detected by measuring the current through it. The term "bolometer" is used for the resistive component as well as the whole detector.

As depicted in Figure 2.17, it is important to isolate the resistive component from its environment thermally so that it retains the energy supplied by the incident photons. Such a suspended structure can be fabricated on a crystalline Si substrate with MEMS technology. For imaging applications, a readout circuit is integrated on the same substrate. An optical cavity is provided between the thin film and the mirror on the substrate. This is to enhance absorption by destructive interference. This feature is useful for setting the wavelength range to be detected efficiently.

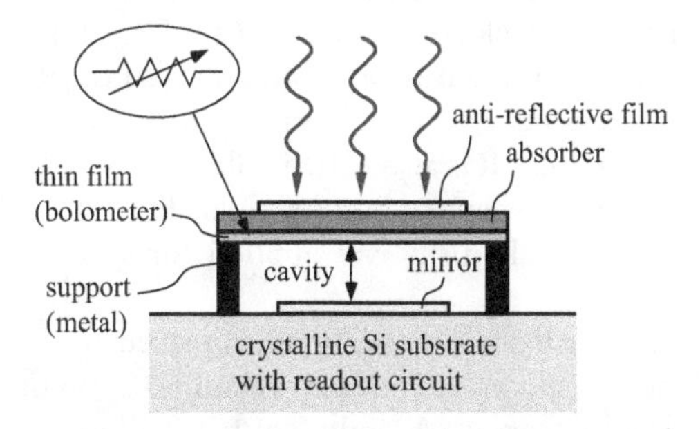

Figure 2.17 Cross-sectional view of a bolometer fabricated on a crystalline Si substrate.

Various metal thin films were investigated as the bolometer material in early studies [36]. Microbolometers have been intensively studied for IR imaging applications. According to the review paper published in 2008, most commercial products adopted either vanadium oxide (VO_x) or amorphous silicon (a-Si) [37].

2.2 Light sources

Light sources are needed for general lighting. Especially useful for many applications are artificial light sources converting electric power to light. They are routinely used to illuminate objects for image acquisition. Various technologies for light emission have been applied to displays. For example, almost all backlight units in transmissive LCDs are based on LEDs.

The mechanisms for light emission are reviewed first and some examples are given. Then, LEDs and OLEDs are described.

2.2.1 Mechanisms of light emission

Physical phenomena utilized by artificial light sources include thermal emission, field emission, and electroluminescence.

Hot bodies emit light when electrons at excited states release their energies as photons. In incandescent lamps and halogen lamps, electric current heats up a filament and it emits light. Plasma is generated by gas discharge and the excited atoms emit light. Neon lamps and sodium lamps utilize this mechanism. Excited atoms also emit ultraviolet photons. Phosphors convert them to visible photons. For example, mercury atoms in a fluorescent tube are excited by gas discharge and they emit ultraviolet photons. Phosphors coated on its interior wall absorb them and emit visible photons. In plasma display panels (PDPs), ultraviolet photons from a xenon gas discharge excite three types of phosphors for generating three primary colors [38]. Down-conversion of photon energy by materials is called photoluminescence. It is called cathodoluminescence when materials are excited by electrons. Three types of phosphors are used in a cathode-ray tube (CRT) [39].

Field emission refers to electron emission under a high electric field in vacuum. Unlike thermionic emission from a heated object, it is a quantum mechanical phenomenon: electrons tunnel through a potential barrier thinned by an external bias. It is also called cold emission. One can accelerate the liberated electrons in vacuum and let them hit a phosphor to generate visible photons. This mechanism was applied for a field emission display (FED) [40].

Vacuum and gas-filled devices are being replaced by solid-state devices. Light emission based on electron–hole pair recombination in a semiconductor is called electroluminescence. A light-emitting diode (LED) is a good example. A series of discoveries and inventions over the last several decades have led to today's success in many applications [41].

2.2.2 Light-emitting diodes

2.2.2.1 Materials and structures

Solving the Schrödinger equation for an electron in a crystal gives the relationship between its energy E and wavevector k. The result is called $E-k$ diagram. Momentum and wavevector are related by the de Broglie's relation: $p = \hbar k$ where $\hbar = h/2\pi$ (Dirac constant). Hence, an $E-k$ diagram allows one to discuss the conservation of momentum for electron–hole recombination.

There are two types of semiconductors: direct and indirect. If the minimum of the conduction band and the maximum energy in the valence band are at the same k in the $E-k$ diagram, the material is called direct semiconductor. When an electron and a hole recombine at this wavevector (i.e., a vertical transition in the diagram), momentum is conserved. Hence, recombination occurs efficiently in a direct semiconductor such as GaAsP alloys. The excess energy associated with this recombination process can be released as a photon. This is not the case for an indirect semiconductor. A third particle (phonon) is needed to conserve momentum for light emission to occur. Unfortunately, crystalline Si is an indirect semiconductor.

Two types of configurations, homo- and heterojunction LEDs, are illustrated in Figure 2.18. In both cases, the top contact is a pad for wire bonding. Because it is opaque, a transparent layer is needed to spread the current laterally. Electrons and holes can recombine in the p-n homojunction diode in Figure 2.18a. When the diode is forward-biased, the external bias

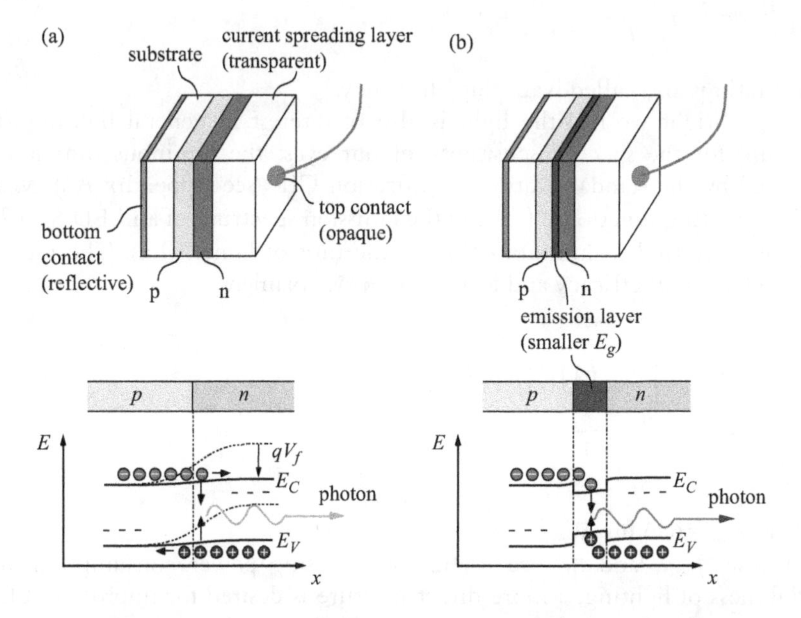

Figure 2.18 Configuration and operation of two types of LEDs under reverse bias: (a) p-n homojunction structure and (b) double heterojunction structure.

V_f reduces the built-in potential by qV_f as illustrated in this figure. Electrons and holes diffuse across the junction more easily and they recombine. The recombination process is more efficient in the double heterojunction structure shown in Figure 2.18b. It is formed by sandwiching a thin semiconductor layer with a larger bandgap material. Because electrons and holes are confined in the central region, recombination proceeds more efficiently.

The bandgap of the emitting material sets the lower limit for the photon energy. Recombination of electrons and holes occupying higher energy states results in emission of a photon with a larger energy. Various inorganic materials have been investigated to cover the visible wavelength range and beyond. For example, AlGaAs and AlInGaN alloys are used for red and blue LEDs, respectively. Wavelengths from red to green are covered by AlInGaP alloys. Controlling the bandgap energies of compound semiconductors by altering their compositions is called bandgap engineering.

2.2.2.2 Efficiency

How efficiently an LED can convert electric power to optical power is of interest. Power efficiency is defined as the ratio of outgoing optical power P_{opt} and electric power supplied from outside. Rewriting the input power as the product of forward bias V_f and forward current I_f, power efficiency is expressed as,

$$\eta_{power} = \frac{P_{opt}}{I_f V_f} \tag{2.13}$$

This ratio is also called wall-plug efficiency.

How bright we feel the light is also of interest in general lighting. To account for the spectral sensitivity of our eyes, the luminous function is defined by the standardization organization CIE (see Appendix A3). With this weighting function $V(\lambda)$ and the emission spectrum of an LED $S_{em}(\lambda)$, P_{opt} is converted to luminous flux in the unit of lumen (lm). The ratio is called luminous efficacy and its unit is lm/W. Namely,

$$\eta_V = \frac{K_m \int_{380}^{780} S_{em}(\lambda) V(\lambda) d\lambda}{I_f V_f} \tag{2.14}$$

where $K_m = 683$ lm/W.

While η_{power} and η_V are useful for accessing power consumption and brightness of lighting, a more direct measure is desired for improving LED characteristics. External quantum efficiency is an index appropriate for this purpose. This is defined as the ratio of the number of photons exiting the

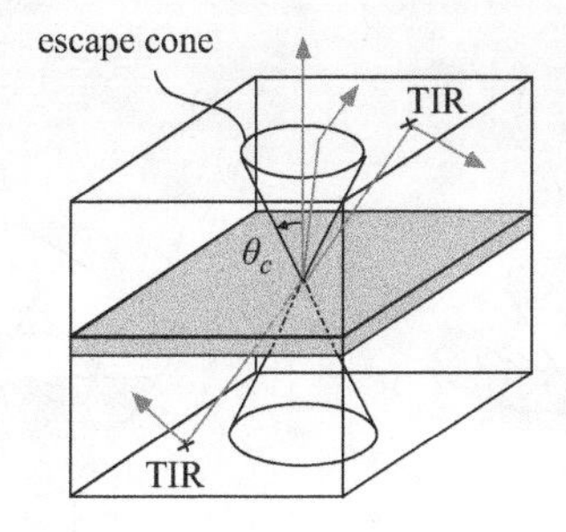

Figure 2.19 Light trapping by TIR. The photons emitted inside the escape cone can exit the device with probabilities given by the Fresnel equations.

device and the number of electrons injected. This is equivalent to the quantum efficiency of a photodiode. For monochromatic photons with energy $h\nu$, the number of photons per unit time exiting the device is equal to $P_{opt}/h\nu$. The number of electrons injected per unit time is expressed as I_f/q where q is the elementary charge. Hence, the external quantum efficiency is given by,

$$\eta_{ext} = \frac{P_{opt}/h\nu}{I_f/q} \tag{2.15}$$

Not all photons can exit the multi-layer structure. As depicted in Figure 2.19, the critical angle for total internal reflection (TIR) is denoted as θ_c. The photons emitted outside the polar angle $\pm\theta_c$ are trapped. Those emitted within $\pm\theta_c$ can exit the structure after refraction. The probability of reflection is given by the Fresnel equations (see Appendix A2). The cone with an apex angle $2\theta_c$ is called an escape cone.

Internal quantum efficiency η_{int} is defined as the ratio of the number of photons generated in the device and the number of electrons injected. Denoting the probability of light extraction as $\eta_{extraction}$, η_{int} and η_{ext} are related by,

$$\eta_{ext} = \eta_{extraction}\eta_{int} \tag{2.16}$$

2.2.2.3 Photon extraction

Efforts to increase $\eta_{extraction}$ continue [42]. For example, surface modifications such as texturing and attaching microstructures extract more photons by breaking the TIR condition. As shown in Figure 2.20, escape cones exist for the side walls of an LED. Shaping an LED to form a truncated inverted pyramid is also effective. The photons reflected by its side walls are directed upward and they break the TIR condition at the top surface.

Current-spreading layer absorbs photons. Bonding pad blocks them. These losses can be avoided by the flip-chip configuration shown in Figure 2.21. The bottom metal electrode is reflective. In this example, the active layer is

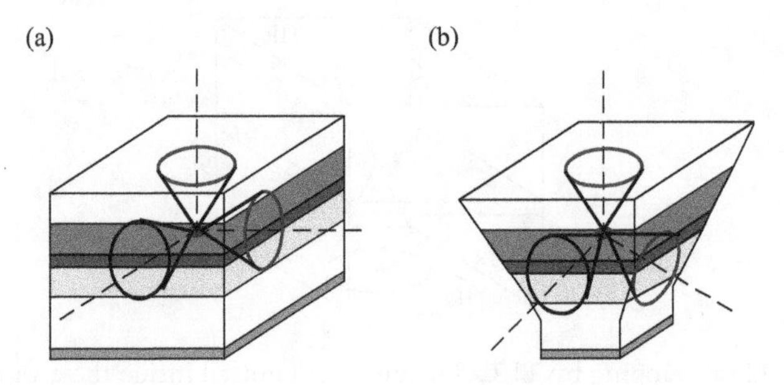

Figure 2.20 Light extraction by shaping an LED die: (a) conventional die and (b) truncated inverted pyramid structure.

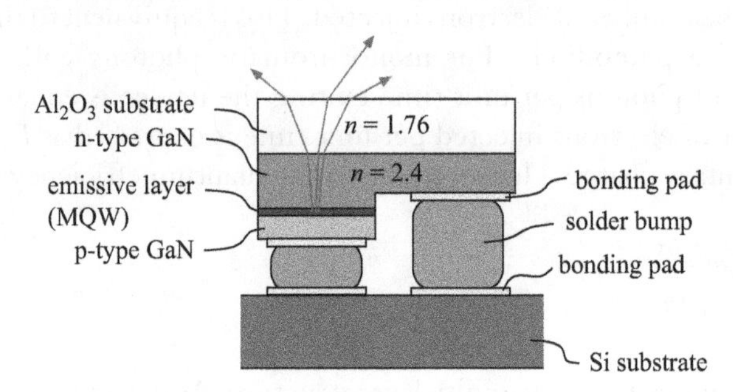

Figure 2.21 Flip-chip bonding allows one to extract light through the substrate.

MQW where multiple heterostructures are formed. Light is directed upward and extracted through the substrate.

2.2.2.4 White LEDs

A white LED that uses phosphors relies on both electroluminescence and photoluminescence [43]. Namely, phosphors are placed around a blue-emitting LED to generate lower-energy photons. The mixture of blue and yellow is perceived as white. Three examples are shown in Figure 2.22. The spectrum labeled as "discrete LED" was measured with a blue LED purchased in 2005. The other two spectra were measured with flashlights built into commercial smartphones in 2022. Although the emission spectra are clearly different, they all look white to our naked eyes. This psychological phenomenon is called metamerism. It is interesting to note that the white LEDs in recent products emit more cyan and green wavelengths (smartphone 1). Having adequate intensity at each wavelength is advantageous for reproducing colors of illuminated objects. Color rendering index is defined to quantify this property of a light source.

Correlated color temperature (CCT) is another index for characterizing a light source. A white LED based on phosphors has a fixed CCT because its

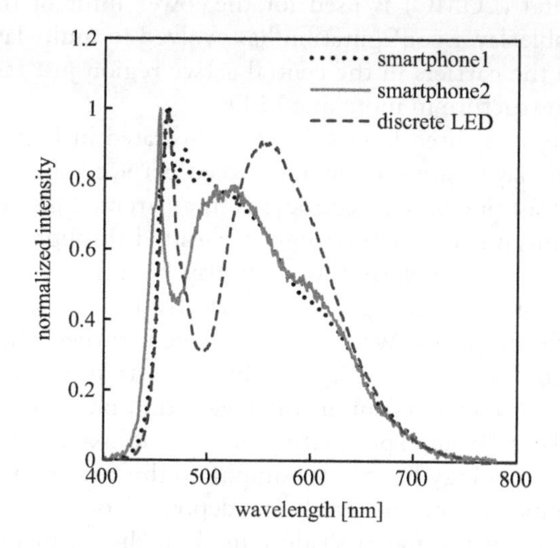

Figure 2.22 Emission spectra of three white LEDs measured in our laboratory.

emission spectrum is fixed. In this regard, simply mixing light from blue, green, and red LEDs can generate white light. By controlling the current in each LED, the CCT of this three-LED system can be tuned.

Furthermore, an interesting configuration for CCT-tunable LEDs was reported in 2020 [44]. Three different MQW structures were vertically stacked with intermediate blocking layers. By applying voltage pulses on this configuration, the emission intensity of each MQW structure was changed, resulting in an adjustable CCT. How this pulsing condition selects each MQW structure to be activated is not clear. But it is interesting to see if this monolithic integration technology can be applied for fabricating practical color micro-displays (see Section 5.3.2).

2.2.3 OLEDs

Organic materials are employed in an organic light-emitting diode (OLED) [45]. Its configuration and operation principle is essentially the same as those of inorganic LEDs. Many consider that the work reported by Tang and VanSlyke [46] was a breakthrough for practical applications. A transparent electrode was formed on a glass substrate, on which two types of organic materials and a metal electrode were formed by vacuum deposition. An electron recombines with a hole in the interface region of the two organic materials, leading to photon emission. This double-layer configuration is equivalent to a p-n diode. Carrier transport in organic materials is also described in a similar manner although somewhat different terms are used. For example, the highest occupied molecular orbital (HOMO) corresponds to the upper limit of the valence band. The lowest unoccupied molecular orbital (LUMO) is used for the lower limit of the conduction band. The double-layer configuration has evolved to multi-layer structures, which confine the carriers in the central active region just as in the double heterojunction structure in inorganic LEDs.

The structure of a three-layer OLED is illustrated in Figure 2.23. Three types of organic layers are stacked and placed between the top and bottom electrodes. At least one of the electrodes is transparent. This device structure is called bottom-emission because light is emitted through the substrate. A top-emission device is preferred for a display because its pixel circuit and metal lines do not block the light. If both electrodes are transparent, light is emitted in both directions. When such a device is turned off, ambient light can go through. This is one way to fabricate a transparent display. Early devices had to be contained in an inert gas atmosphere with desiccant to prevent degradation from exposure to moisture and oxygen. Later, thin-film encapsulation technology has greatly simplified this packaging issue [47,48].

Small-molecule organic materials are deposited on a large substrate by vacuum evaporation through a shadow mask as shown in Figure 2.24a. A very large shadow mask can be deformed by its own weight. Molecules moving around the edges of the mask can degrade the accuracy of deposition.

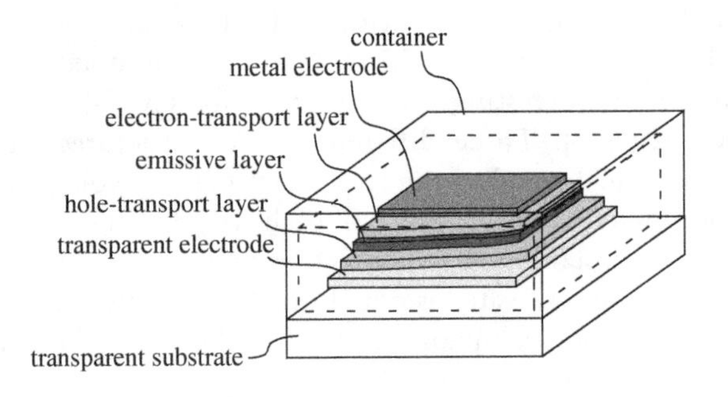

Figure 2.23 Early packaging for a three-layer OLED.

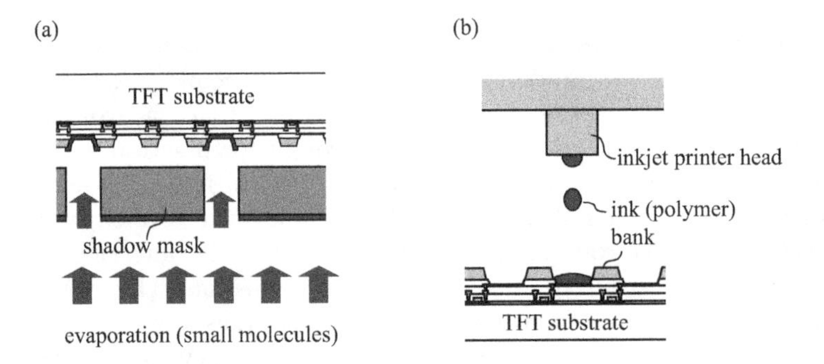

Figure 2.24 Deposition of organic layers over a large area: (a) evaporation through a shadow mask and (b) inkjet printing.

Polymer materials can be deposited by an inkjet printer. Prior to printing, hydrophobic banks are formed on the substrate to prevent the droplet from spreading as shown in Figure 2.24b. In both cases, OLEDs are well suited for large-area applications such as direct-view displays and solid-state lighting. Additive color mixing is realized by placing three types of OLEDs emitting each primary color side by side. Alternatively, either color-conversion materials are stacked on blue-emitting OLEDs, or color filters are stacked on white-emitting OLEDs.

2.3 Transistors

A transistor is a three-terminal device. A metal–oxide–semiconductor field-effect transistor (MOSFET) is by far the most common transistor used in digital and analog circuits. In fact, one fabricated on a crystalline Si substrate (wafer) is the building block of an integrated circuit (IC). Both n-type and p-type MOSFETs are fabricated side by side. A complementary MOS (CMOS) circuit is formed by connecting them. An inverter is a simple example of a CMOS circuit. As shown in Figure 2.25, p-type and n-type MOSFETs are connected in series and an input voltage is applied to their gate terminals. When the input voltage is high enough, the n-type MOSFET is turned on. The output terminal is connected to the ground through its on-time resistance. The logic is inverted. Because one of the MOSFETs is always turned off, no steady current flows through them. Thus, power consumption of a CMOS circuit is low.

Because insulators other than oxides are also used, more general terms such as metal insulator field-effect transistor (MISFET) and insulated gate field-effect transistor (IGFET) should be used in principle. In addition, a standard gate material is a heavily doped polycrystalline silicon (poly-Si) rather than a metal. Despite these facts, the term MOSFET is commonly used to refer to this type of transistor.

For large-area electronics applications, one can fabricate transistors on one substrate and transfer them to another substrate. For example, single-crystalline Si MOSFETs fabricated on a Si wafer were transferred to a glass substrate [49]. Polycrystalline Si TFTs fabricated on a glass substrate were transferred to a polymer substrate for flexible device applications [50]. While these transfer

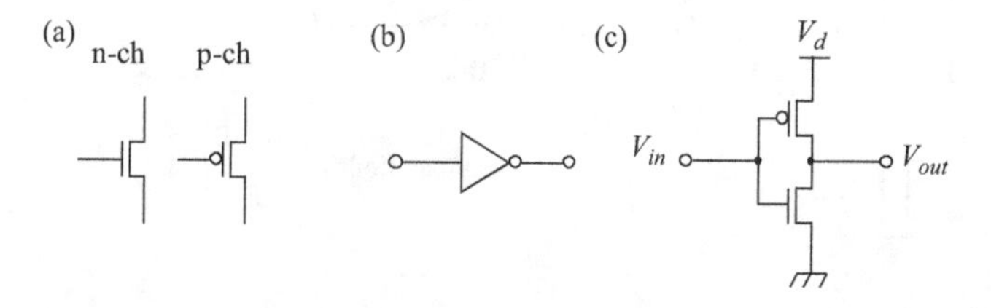

Figure 2.25 A simple example of a CMOS circuit: inverter: (a) symbols for MOSFETs, (b) a symbol for an inverter, and (c) its CMOS circuit.

techniques aim to take advantage of superior characteristics of individual transistors, they bear the burden of process complexity.

In practical applications, thin-film transistors are fabricated directly on glass substrates at low temperatures in most cases. Depending on the semiconductor materials, they are called a-Si TFTs, poly-Si TFTs, metal-oxide TFTs, etc. Their electrodes are much larger than those of crystalline Si MOSFETs. Metals are used for the transistor terminals and for connecting TFTs and other circuit components. For driving a large-area image sensor and a display, one needs to be careful about the analog nature of a circuit. For example, each pixel of a display needs to be addressed fully in an allocated time duration. An unintended capacitance is formed between each terminal of a TFT. These capacitances are charged and discharged during the addressing operation. The feedthrough charge from the gate terminal results in fixed-pattern noise (FPN) in an image sensor (see Section 4.1.3).

In this section, the basic theory for a MOSFET is reviewed first and descriptions of various TFTs follow.

2.3.1 MOSFETs

2.3.1.1 MOS structure

Let us consider the behavior of charge carriers in a metal–oxide–semiconductor (MOS) structure. The following knowledge is also required for understanding a CCD. A band diagram of a MOS structure for a p-type Si is illustrated in Figure 2.26. The vacuum level and the work function of metal and Si are denoted as E_{vac}, ψ_M, and ψ_{Si}, respectively. Flat-band voltage V_{fb} is defined as the voltage applied to the metal such that the Fermi level in the metal coincides with the Fermi level in the semiconductor as shown in Figure 2.26a.

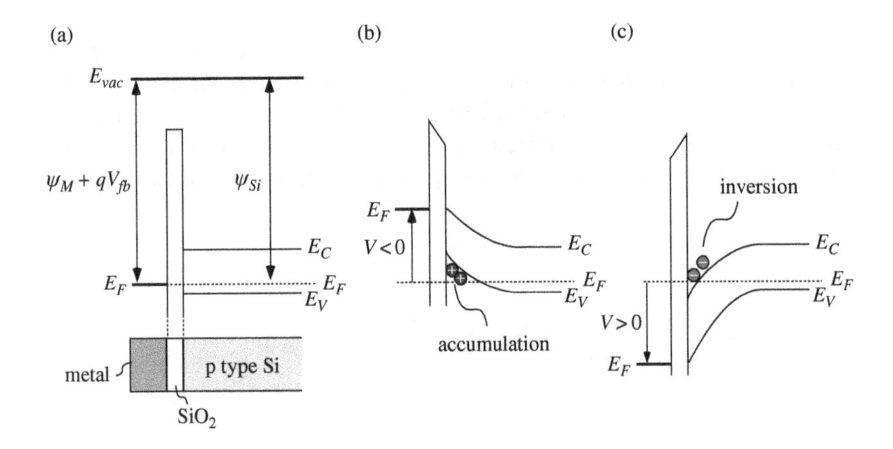

Figure 2.26 MOS structure formed on a p-type Si wafer and band diagrams corresponding to three biasing conditions: (a) flat-band, (b) accumulation, and (c) inversion.

When a negative bias is applied to the metal, the band near the insulator bends upward as shown in Figure 2.26b. The majority carrier in the p-type Si, namely, holes, accumulate at the region near the insulator. If the bias applied to the metal is positive, the band bends downward. The holes move away from this region. As the bias is increased, the band bends further downward. When it exceeds a certain level called threshold voltage V_{th}, electrons start to accumulate in this surface region as shown in Figure 2.26c. Although the majority carrier in a p-type semiconductor is a hole, an electron-rich region appears. This region is called an inversion layer.

Under the conditions of accumulation and inversion, the charge carriers in the surface region can move along the direction perpendicular to the cross section if an appropriate electric field exists. This effect is utilized in a MOSFET as well as in a CCD.

2.3.1.2 Configuration

Cross section of MOSFETs on a doped crystalline Si wafer is shown in Figure 2.27. The left-hand side is an n-channel MOSFET and the other one is a p-channel MOSFET. They are separated by a trench filled with SiO_2 called shallow-trench isolator. It is needed to prevent the unintended formation of p-n junctions and other anomalies that might arise in circuit designs.

A MOSFET is essentially a stack of three layers (a metal, an insulator, and a semiconductor) with p-n junctions formed on both ends. The metal and the insulator are called the gate, and gate insulator, respectively. The bias applied to the gate can induce charges at the surface of the semiconductor. They can flow between the two terminals called source and drain if a lateral field exists. This current is controlled by the gate bias V_g and the drain bias V_d. The source is usually grounded.

2.3.1.3 Current-voltage characteristics

Let us consider an n-channel MOSFET illustrated in Figure 2.28. When the gate bias V_g exceeds the threshold voltage V_{th}, charges appear under the gate oxide. This is called channel. Its length and width are denoted as L and W,

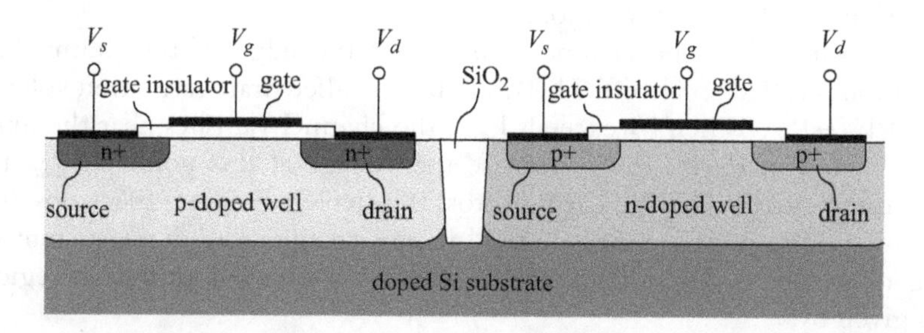

Figure 2.27 Cross section of an n-channel MOSFET and a p-channel MOSFET isolated by a trench filled with SiO_2.

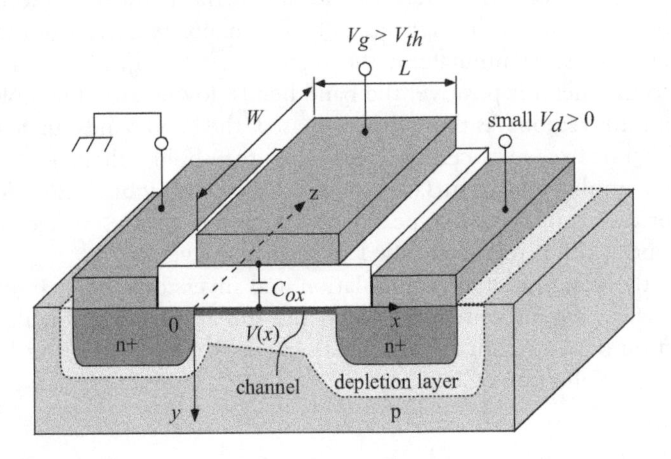

Figure 2.28 Geometry of an n-channel MOSFET.

respectively. A capacitance is formed between the channel and the gate. This capacitance per unit area is denoted as C_{ox}. Note that its unit is F/cm^2.

Field-effect mobility is denoted as μ_{fe}. Note that μ_{fe} can be different from the mobility in the bulk of a semiconductor. Charge carriers need to flow in the region near the interface. Any irregularities such as dangling bonds and impurities can scatter the electrons [51]. The following expressions for the drain current are derived under the assumption called gradual channel approximation (see Appendix A1.2). They are valid for a long-channel device.

Suppose that the source is grounded and that a positive bias V_d is applied to the drain. For $V_d \le V_g - V_{th}$, the channel extends from the source to the drain. The drain current I_d flows from the drain to the source. It is expressed as follows.

$$I_d = \mu_{fe} C_{ox} \frac{W}{L} \left\{ \left(V_g - V_{th} \right) V_d - \frac{V_d^2}{2} \right\}, \quad V_d \le V_g - V_{th} \tag{2.17}$$

If V_d is very small, the second term in the parenthesis is neglected. Then, I_d is proportional to V_d. Thus, the transistor behaves like a resistor. This mode of operation is called a linear region.

At $V_d = V_g - V_{\text{th}}$, the channel vanishes at the edge of the drain. This condition is called pinch-off. This drain bias is called drain saturation voltage: $V_{dsat} \equiv V_g - V_{\text{th}}$. When V_d exceeds V_{dsat}, the channel vanishes near the drain as illustrated in Figure 2.29. Because the voltage at this point is V_{dsat}, the extra drain voltage $V_d - V_{dsat}$ drops across this depleted region. Electrons drift toward the drain by the electric field in this region. The drain current no longer depends on V_d. This operation condition is called saturation region. It is given by,

$$I_d = -\mu_{fe} C_{\text{ox}} \frac{W}{L} \frac{\left(V_g - V_{th} \right)^2}{2}, \qquad V_g - V_{th} < V_d \tag{2.18}$$

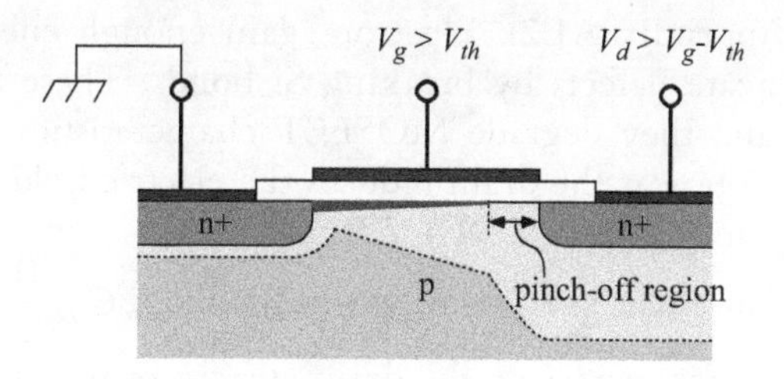

Figure 2.29 For $V_d > V_g - V_{th}$, the channel region near the drain is depleted. In this pinch-off region, a high electric field exists.

Example 2.5

A MOSFET has the following parameters: $V_{th} = 2.0\text{V}$, $W/L = 200$. Electron mobility $\mu_e = 100\ \text{cm}^2/\text{V}\cdot\text{s}$.

1. When this MOSFET is operated at $V_g = 10\text{V}$ and $V_d = 4.0\text{V}$, its drain current is 960 μA. Calculate the oxide capacitance per unit area.
2. The drain bias of this MOSFET is increased to 10 V. Calculate its drain current.

Solution

1. Because $V_g - V_{th} = 8.0\text{V} > V_d$, the MOSFET is in the linear region. Hence,

$$C_{ox} = \frac{I_d}{\mu \frac{W}{L}\left[\left(V_g - V_{th}\right)V_d - \frac{V_d^2}{2}\right]}$$

$$= \frac{960 \times 10^{-6}\,\text{A}}{100\,\text{cm}^2/\text{Vs} \times 200 \times \left(8.0\,\text{V} \times 4.0\,\text{V} - \frac{4.0^2\,\text{V}^2}{2}\right)}$$

$$= 2.0 \times 10^{-9}\,\frac{\text{A}\cdot\text{s}}{\text{cm}^2\cdot\text{V}} = 2.0 \times 10^{-9}\,\text{F/cm}^2$$

2. Because $V_g - V_{th} < V_d$, the MOSFET is in the saturation region. Hence,

$$I_d = \mu C_{ox}\frac{W}{L}\frac{\left(V_g - V_{th}\right)^2}{2}$$

$$= 100\text{cm}^2/\text{Vs} \times 2.0 \times 10^{-9}\,\text{F/cm}^2 \times 200 \times \frac{8^2\,\text{V}^2}{2}$$

$$= 128 \times 10^{-5}\,\frac{\text{F}\cdot\text{V}}{\text{s}} = 1.28 \times 10^{-3}\,\text{A}$$

Under the pinch-off condition, the electric field near the drain becomes very high (see Appendix A1.2). Electrons gain enough energy for impact ionization and create defects by breaking Si bonds. These are called hot-electron effects and they degrade MOSFET characteristics [52]. Adding a lightly doped region near the drain reduces the electric field. This structure is called a lightly doped drain (LDD) [53].

Drain current of a MOSFET is proportional to $\mu_{fe} C_{ox} \frac{W}{L}$. Hence, there are three strategies for increasing the drain current or making the transistor area smaller for larger-scale integration. First, reducing L has been the route taken so far to sustain Moor's law. The gradual channel approximation is applicable to a long-channel device. For a short-channel device, so-called short-channel effects must be considered. Second, increasing C_{ox} is also effective. This requires a thinner oxide layer or the use of an insulator with a higher dielectric constant [54]. Third, replacing Si with other semiconductor materials such as strained Si [55,56] can enhance μ_{fe}. Note that the same arguments apply to TFTs.

2.3.2 Thin-film transistors

2.3.2.1 Overview

An amorphous silicon (a-Si) TFT was proposed by P. G. Lecomber et al. in 1979. The potential for driving LCDs with a-Si TFTs was demonstrated soon after [57,58]. Extensive research and development activities worldwide resulted in some commercial products of LCDs in the 1980s [59]. During the next two decades, a-Si TFT technology dominated the flat-panel display industry, especially for large-size LCDs. Note that a-Si TFTs were used only for addressing each pixel. Electron mobility of about 1 $cm^2/V \cdot s$ was considered adequate for this purpose. ICs had to be mounted at the peripheral region of a glass substrate. Thousands of metal lines needed to be connected to supply signals for each pixel. Enlarging a glass substrate was an effective tactic for cost reduction. It reached about 10 m^2 around 2010. Exact dimensions of glass substrates can vary among the same generation of production facilities. One example of so-called Gen 10 facilities is reported to use 2,850 mm × 3,050 mm substrates [60].

In the mid-1980s, low-temperature polycrystalline silicon (LTPS) TFTs were fabricated on an inexpensive glass substrate by a process called excimer laser annealing (ELA) [61]. Since then, they have been investigated mainly for applications in small to mid-size LCDs [62]. Mobility of electron and hole exceeds 100 and 50 $cm^2/V \cdot s$, respectively. Because CMOS circuits can be integrated on a glass substrate with LTPS TFTs, external driver ICs are no longer needed. This unique feature remains the advantage of LTPS technology. Since around 2000, LTPS TFTs have been also used to drive

OLED displays [63]. However, equipment cost associated with ELA limits the size of substrates.

In 2004, n-channel TFTs were fabricated with amorphous indium gallium zinc oxide (a-IGZO) on a plastic substrate at room temperature. The field-effect mobility was about $6\text{cm}^2/\text{V}\cdot\text{s}$–$9\text{cm}^2/\text{V}\cdot\text{s}$ [64]. Because active layers are deposited by sputtering, most of the manufacturing facilities for a-Si TFTs can be used to fabricate a-IGZO TFTs. The number of publications on a-IGZO TFTs declined slightly in 2010, indicating that commercialization was imminent [65]. In the mid-2010s, the display industry gradually accepted a-IGZO TFTs for driving large-size LCDs and OLED displays [66]. However, attempts to fabricate p-channel TFTs with characteristics comparable to n-channel TFTs have been challenging [67].

In addition, organic TFTs were widely studied for display and sensor applications. They can be fabricated on a flexible substrate by solution processes at low temperatures [68]. Printing TFTs in a roll-to-roll process would simplify the fabrication process [69]. Field-effect mobility as high as 31.3 $\text{cm}^2/\text{V}\cdot\text{s}$ was measured for single-crystal organic TFTs fabricated by inkjet printing [70]. A theoretical study suggested that hole mobility in such materials can reach some tens of $\text{cm}^2/\text{V}\cdot\text{s}$ [71]. Most organic TFTs are p-channel devices, but the possibility of CMOS circuits exists. While these mobility values are impressive, the display industry is slow to adopt organic TFTs because of existing competitive TFT technologies. Organic TFTs might find applications in low-cost devices such as disposable sensors [72] and radio frequency identification systems [73].

Note that all TFTs are MOSFETs. In general, a TFT controls the charges accumulated in its channel. In contrast, a crystalline Si MOSFET controls the inverted charges at the channel. In other words, a TFT is operated under accumulation mode while a crystalline Si MOSFET is operated under inversion mode.

2.3.2.2 Amorphous Si (a-Si) TFTs

A typical structure of an a-Si TFT is illustrated in Figure 2.30. A gate electrode, a gate insulator, and an a-Si layer are stacked in this order. This configuration is called inverted staggered because the stacking sequence of these layers is inverted from its original design. It is also called bottom-gate structure. There are two additional features not shown in the figure. First, an n+ a-Si layer is inserted between the a-Si layer and the metal to make the contacts ohmic. Second, the upper surface of the a-Si layer is etched slightly to make it inactive. This additional process is called backchannel etch.

Non-alkali glass substrates are used to fabricate TFTs to avoid complications arising from alkali elements. They are heated to about 350 °C for depositing a-Si films by PECVD (see Section 3.1.1). This corresponds to the maximum process temperature for fabricating a-Si TFTs. The gate insulator material is amorphous silicon nitride (a-SiN_x:H) prepared by PECVD. A metal layer

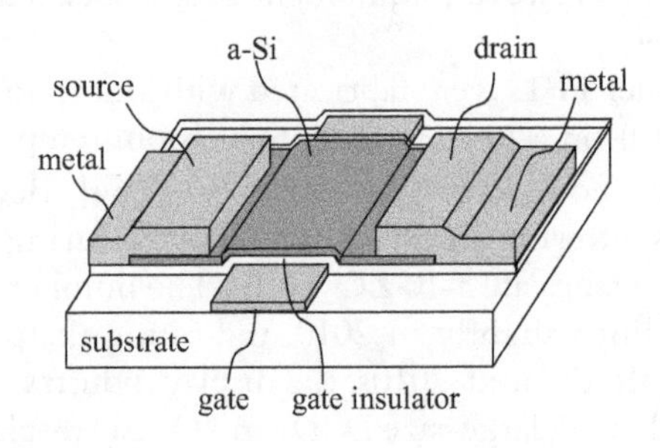

Figure 2.30 Structure of an inverted-staggered a-Si TFT.

is deposited and patterned by photolithography to form source and drain electrodes.

Electron mobility in a-Si is about $0.5 \sim 1\,\text{cm}^2/\text{V}\cdot\text{s}$. This is adequate to address each pixel in a large-size LCD. It is an n-channel device. Because the hole mobility in a-Si:H is smaller by two orders of magnitude, a p-channel device is not practical. Therefore, no CMOS circuits can be integrated with a-Si TFTs. Besides, threshold voltage shifts during a prolonged bias stress. It is widely accepted that the mechanisms behind the threshold voltage shift are defect creation in an a-Si:H layer or at the a-Si:H / a-SiN$_x$:H interface and charge trapping in the gate insulator. The hydrogen concentration in a-SiN$_x$:H is related to the field-effect mobility and TFTs fabricated with N-poor gate insulator show hysteresis in current-voltage characteristics due to charge trapping at the interface [74].

The threshold voltage shift under bias stress poses a problem for analog operation of TFTs [75]. For example, in an OLED display, current flows in each OLED almost continuously. The threshold voltage shift in the TFT connected in series of the OLED results in a luminance drop.

2.3.2.3 LTPS TFTs

There are two techniques to fabricate poly-Si films from a-Si films: solid phase crystallization (SPC) and laser annealing. These are called high- and low-temperature processes, respectively. Because SPC requires a temperature higher than 900°C, an expensive quartz substrate is needed. High-temperature poly-Si films are applied for liquid crystal light valves in a projector (see Section 5.5.1). To avoid the deformation of glass substrates, the maximum process temperature must be below 600°C. Irradiating an a-Si film with a pulsed laser light crystallizes it. Although the surface temperature exceeds the melting point of Si (1,410°C), there is no thermal damage to the substrate. This crystallization process is called laser annealing (see Section 3.2).

Configuration of a LTPS TFT is coplanar, i.e., the three terminals are on one side of the channel. Its fabrication process steps are illustrated in Figure 2.31. First, an a-Si layer is deposited on a glass substrate by low-pressure chemical vapor deposition (LPCVD). It is converted to a poly-Si film by laser annealing. The poly-Si film is patterned by photolithography such that the remaining poly-Si layer becomes the source, drain, and active region of a TFT. To control threshold voltage, impurity atoms are introduced. As shown in Figure 2.31a, a SiO_2 layer is deposited on this poly-Si region. It serves as a gate insulator. Next, a metal layer is deposited and patterned as shown in Figure 2.31b. It becomes a gate electrode. Next, impurity atoms are introduced by ion implantation. The patterned metal serves as a mask and blocks ions. This ensures that impurity atoms are introduced to the semiconductor regions very close to the gate as shown in Figure 2.31c. This concept is called self-alignment. Equipment for ion implantation without mass separation is called ion shower. Although impurity atoms can be introduced over a large area, their process control is reported to be challenging. The substrate is heated so that the introduced impurity atoms are incorporated into the Si network. This process is called dopant activation. Finally, as shown in Figure 2.31d, an oxide is deposited to serve as an interlayer insulator. Contact holes are etched through it and metal wirings are created to connect the source and drain to other components.

Field-effect mobility of LTPS TFTs exceeds $400\,\mathrm{cm^2/V \cdot s}$ for electrons and $100\,\mathrm{cm^2/V \cdot s}$ for holes [76]. For large-area applications, it is critically important to have uniform characteristics. Also important is the parasitic capacitance between the gate and the source/drain. A large parasitic capacitance would increase the time duration and power consumption required for addressing each pixel. Self-alignment process for TFTs with coplanar configuration reduces parasitic capacitance by minimizing the overlapping areas between the gate and the source/drain.

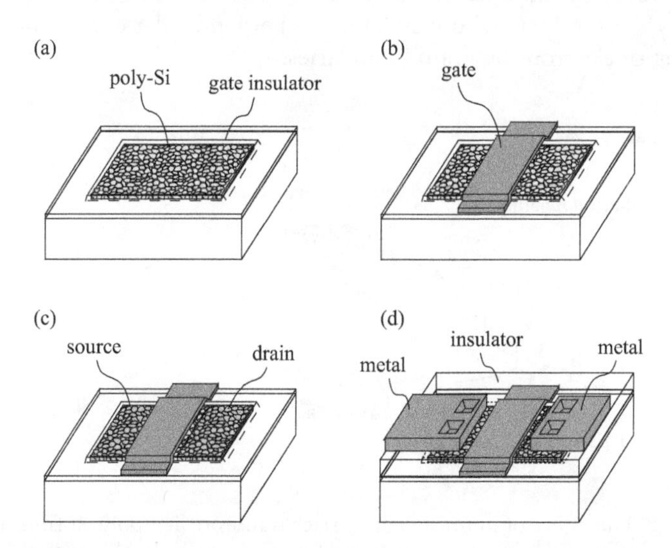

Figure 2.31 Fabrication steps for poly-Si TFTs.

Field-effect mobility of LTPS TFTs increases as poly-Si grains become larger [77]. Because there are dangling bonds at grain boundaries, they hinder carrier transport by blocking and/or trapping them. However, a large grain is not necessarily beneficial for display and sensor applications. In a film with large grains, the number of grain boundaries between the source and drain of a TFT is small. Carriers encounter grain boundaries only occasionally as schematically drawn in Figure 2.32b. When the number is small, its variation is large. Hence, the relative variation in field-effect mobility increases with grain size as shown in Figure 2.32a. In displays and image sensors, millions of TFTs need to behave in a similar manner. Hence, a practical design would sacrifice field-effect mobility over uniformity. Grain size can be controlled by laser annealing conditions (see Section 3.2).

Before excimer laser annealing was actively investigated in the mid-1980s, the concept of lateral growth was demonstrated with continuous-wave (CW) lasers. Since around 2000, there has been renewed interest in CW laser crystallization. A possible motivation behind this is the high maintenance cost of gas-based excimer lasers.

In 2002, a diode-pumped solid-state (DPSS) laser (Nd:YVO_4) emitting at 532 nm was used to fabricate large-grain poly-Si TFTs on 300 mm × 300 mm non-alkali glass substrates [78]. First, a 150 nm-thick a-Si film was patterned to form 50 µm × 200 µm islands. They were crystallized by scanning a laser beam with a spot size of 400 µm × 20 µm without an encapsulation layer. Nucleation started at the edges of the patterned a-Si films. Large grains of about 3 µm × 20 µm were obtained at the central regions of the islands. Grain boundaries were mostly parallel to the scanning direction. The average field-effect mobility of electrons was 422 $cm^2/V \cdot s$ for the n-channel TFTs whose current direction was along the scanning direction. It was 290 $cm^2/V \cdot s$ for TFTs whose current was perpendicular to it, indicating the scattering of electrons by grain boundaries.

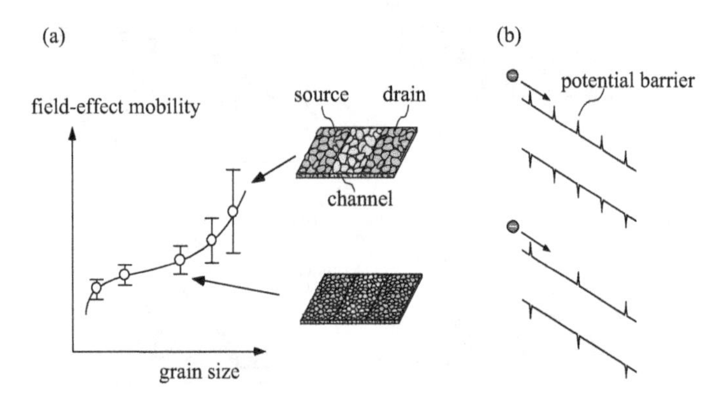

Figure 2.32 The effect of grain size on carrier transport in a poly-Si film. (a) Field-effect mobility as a function of grain size. (b) In the energy band diagram, potential barriers and traps are represented by kinks and levels in the bandgap, respectively.

In 2010, multiple semiconductor laser diodes emitting at 445 nm were used to crystallize 50 nm-thick a-Si films prepared by PECVD on glass substrates. It resulted in films with uniform micro-grains [79]. In 2021, the field-effect mobility of 169 $cm^2/V \cdot s$ was reported for LTPS TFTs on polyimide substrates [80]. Blue semiconductor laser diodes emitting at 445 nm were used to crystallize a-Si films prepared by PECVD. The average grain size of the poly-Si films was wider than 3 μm and longer than 10 μm.

2.3.2.4 *Single-grain TFTs*

If there are no grain boundaries in a channel region, it is a single-crystal Si TFT. Such transistors were reported in 1982 [81]. First, a poly-Si film was deposited on a fused silica substrate. The film was patterned to form 50 μm × 20 μm islands and they were encapsulated with silicon dioxide. These poly-Si islands were recrystallized by scanning a Gaussian-shaped CO_2 laser beam. Note that this was a CW laser. Transistors with a channel length of 12 μm and a width of 20 μm were fabricated and the field-effect mobility of these TFTs exceeded 1,000 $cm^2/V \cdot s$. Although fused silica substrates are expensive for large-area applications, this work clearly indicates a promising future for the CW laser crystallization technique.

2.3.2.5 *Metal-oxide TFTs*

Structure of a-IGZO TFTs is the same as that of a-Si TFTs, namely, inverted staggered configuration (also called bottom-gate structure). Metal-oxide layers are usually deposited by sputtering. Hence, one can replace PECVD in production facilities of a-Si TFTs by sputtering to manufacture a-IGZO TFTs. Electron mobility of an a-IGZO TFT is of the order of 10 $cm^2/V \cdot s$. Due to its low leakage current, its on-off ratio reaches 10^8. This is much higher than those of a-Si TFTs and LTPS TFTs.

Because the drain current of a TFT is proportional to $\mu_{fe} C_{ox} \frac{W}{L}$, one wishes to increase μ_{fe} and decrease L. Coplanar configuration (also called top-gate structure) is advantageous for making a short-channel transistor because of the self-alignment process. Parasitic capacitances between the gate and the source/drain also decrease, resulting in faster pixel addressing and a reduction of power consumption. New materials might increase μ_{fe}. These approaches apply to metal-oxide TFTs as well. In 2009, top-gate TFTs based on amorphous indium zinc oxide (a-InZnO) were reported to have μ_{fe} of 115 $cm^2/V \cdot s$ [82].

Unfortunately, it is difficult to fabricate p-channel TFTs with comparable characteristics to n-channel TFTs. This has led to the concept of designing CMOS circuits with n-channel a-IGZO TFTs and p-channel either LTPS TFTs [83,84] or organic TFTs [85].

What's unique about metal-oxide TFTs is transparency because of the wide bandgap of 3.2 eV for a-InGaZnO. Transparent OLED displays present an interesting prospect for augmented reality applications [86].

2.4 Wavefront-control devices

Image acquisition and display systems manipulate photons in a variety of ways. They utilize various wavefront-control devices. For example, microlens arrays are used to focus incoming light on pixels in an image sensor. Optical films in a backlight unit control the direction of light propagation. Circular polarizers improve the image quality of OLED displays by reducing the reflection of ambient light. These are all passive devices in the sense that their optical properties are fixed. In contrast, active devices are realized by liquid crystal and MEMS technologies. For example, an external electric field alters the orientations of liquid crystal molecules at each pixel in LCDs. This results in a change in the transmittance or reflectance of a pixel (see Section 5.1). Coulomb force moves small mirrors in each pixel of a DMD. This active control is adopted by projectors (see Section 5.5).

In this subsection, control of polarization states with passive devices is described first. It serves as a base for understanding liquid crystal devices. Then, active wavefront control by liquid crystal and MEMS technologies are described.

2.4.1 Passive devices

Polarizers and waveplates alter polarization states of the light passing through them. One can modulate its intensity by combining these passive devices. Jones vectors are used to express polarization states. Jones matrices change polarization states by acting on Jones vectors. An introduction to Jones calculus is found in Appendix A2.2.

2.4.1.1 Polarizers

Suppose that unpolarized light propagates along the z axis. It is decomposed into two polarization components: one with its electric field along the x axis and the other along the y axis. As illustrated in Figure 2.33, a linear polarizer transmits one component only. The y axis in this case is called transmission axis. The other component is either absorbed or reflected. Because this polarizer transmits vertical components, its Jones matrix is $\begin{bmatrix} 0 & 0 \\ 0 & 1 \end{bmatrix}$.

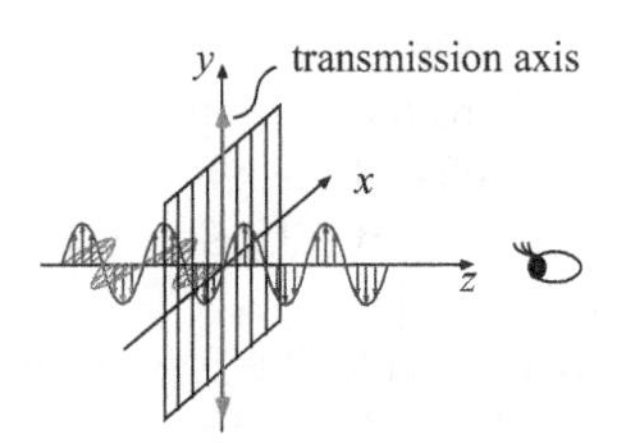

Figure 2.33 Linearly polarized light emerges from a linear polarizer.

Absorptive polarizers are commonly used in LCDs. A polymer material, polyvinyl alcohol (PVA) is doped with iodine. When it is stretched, electrons in iodine can move more easily along the stretched direction. Therefore, the light polarized along the polymer chain transfers its energy to the electrons more easily. Namely, it is preferentially absorbed. For protecting PVA, triacetyl cellulose (TAC), an optically isotropic material developed originally for photographic films, is used. The thickness of a linear polarizer is of the order of 100 μm and two of them are required for a transmissive LCD.

In-cell polarizer technology incorporates the function of a linear polarizer inside a liquid crystal cell [87]. Protection films are no longer needed, resulting in a much thinner display based on liquid crystal technology. In-cell polarizers might also be applied for increasing the contrast ratio of LCDs by eliminating scattered light inside a liquid crystal cell [88].

2.4.1.2 *Waveplates*

A waveplate, also called phase retarder, is made of a birefringent material. Suppose that linearly polarized light is normally incident on a waveplate of thickness d as illustrated in Figure 2.34. Let us denote its extraordinary and ordinary refraction indices as n_e and n_o, respectively. It has slow and fast axes.

For the polarization component along its slow axis, the waveplate exhibits a refractive index n_e. This component propagates at c/n_e inside the waveplate. For the component along the fast axis, the waveplate exhibits n_o and its velocity is c/n_o. The difference in optical distances between the two components is $(n_e - n_o)d$. The phase difference between them is given by,

$$\Delta = \frac{2\pi}{\lambda}(n_e - n_o)d \tag{2.19}$$

Because the phase of the component polarized along the slow axis is delayed by Δ, the Jones matrix of a waveplate is expressed as,

$$W(\Delta) = \begin{bmatrix} 1 & 0 \\ 0 & e^{-i\Delta} \end{bmatrix} \tag{2.20}$$

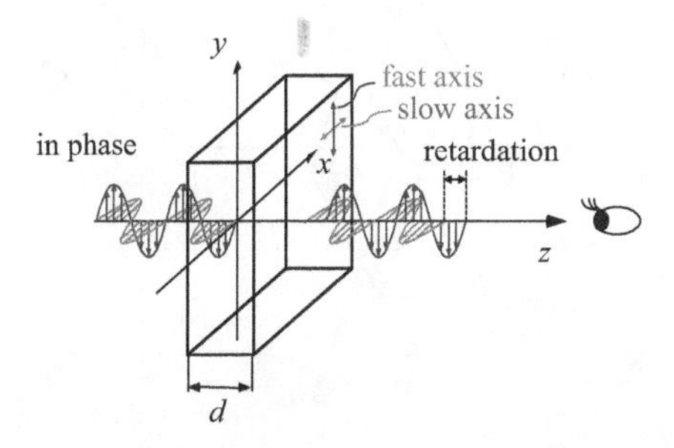

Figure 2.34 A waveplate with its slow axis along the horizontal direction. The phase of the polarization component along the horizontal axis is retarded by Δ.

A waveplate with $\Delta = \pi/2$ is called a quarter-waveplate (QWP). If $\Delta = \pi$, it is called half-waveplate (HWP).

2.4.1.3 Modulation of transmittance

Let us consider the configuration illustrated in Figure 2.35. A waveplate is sandwiched by two crossed linear polarizers. The optical axis of the waveplate is rotated by the angle θ as defined in the figure.

To account for rotation of the coordinate system, the following Jones matrix is needed.

$$R(\theta) = \begin{bmatrix} \cos\theta & \sin\theta \\ -\sin\theta & \cos\theta \end{bmatrix} \tag{2.21}$$

The Jones matrix for the rotated waveplate is given by the product of each Jones matrix, i.e., $R(-\theta)W(\Delta)R(\theta)$.

Suppose that unpolarized light is incident on the linear polarizer on the left. Linearly polarized light passes through the waveplate. Jones vector of the light exiting the polarizer on the right is calculated as follows.

$$\begin{bmatrix} 1 & 0 \\ 0 & 0 \end{bmatrix} R(-\theta)W(\Delta)R(\theta) \begin{bmatrix} 0 \\ 1 \end{bmatrix} = \frac{1}{2}\begin{bmatrix} \left(1-e^{-i\Delta}\right)\sin 2\theta \\ 0 \end{bmatrix} \tag{2.22}$$

Finally, the transmittance of this light is given by,

$$T = \left|\frac{1}{2}\left(1-e^{-i\Delta}\right)\sin 2\theta\right|^2 = \sin^2 2\theta \cdot \sin^2 \frac{\Delta}{2} \tag{2.23}$$

If $\Delta = \pi$ (i.e., for the case of an HWP), the equation above reduces to $I = \sin^2 2\theta$. Hence, T varies from 0 to 1 when the HWP is rotated by 45°.

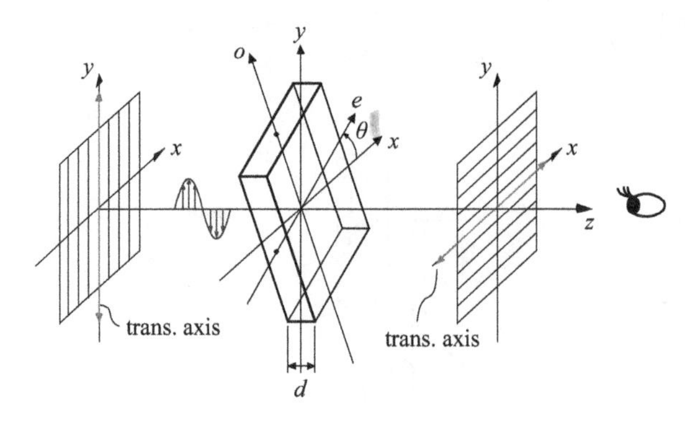

Figure 2.35 Two crossed polarizers sandwiching a rotated waveplate.

Alternatively, one can set θ to 45° and the equation above is reduced to $I = \sin^2 \frac{\Delta}{2}$. By changing Δ electrically, T can be set from 0 to 1. Liquid crystal devices utilize these facts to modulate light intensity.

In addition, a waveplate sandwiched by crossed polarizers constitutes the foundation of a polarized optical microscope. It identifies anisotropic features in specimens. For example, grains in a polycrystalline organic semiconductor film are clearly observed under a polarized optical microscope [89,90].

Example 2.6

Unpolarized monochromatic light at wavelength 550 nm is incident on an HWP waveplate sandwiched by crossed polarizers. The refractive index of this waveplate varies from 1.50 to 1.65. Calculate the minimum thickness required for this waveplate to modulate the transmittance of this light from 0 to 1.

Solution

Phase retardation of a HWP is given by $\pi = \frac{2\pi}{\lambda}(n_e - n_o)d$. $\therefore d = \frac{\lambda}{2(n_e - n_o)} = \frac{550\,\text{nm}}{2(1.65 - 1.50)} = 1.83\cdots\mu\text{m}$. Hence, the answer is 1.8 μm. Note that the number of reliable digits decreases to two in the subtraction process.

2.4.1.4 *Modulation of reflectance*

Because organic layers in an OLED are almost transparent to visible light, incoming ambient light can be reflected by its metallic electrode. If untreated, the contrast ratio of OLED displays decreases under a bright environment. Placing a circular polarizer on an OLED display can prevent this.

Let us consider stacking a linear polarizer on a QWP. The angle between the transmission axis of the linear polarizer and the slow axis of the QWP is set to 45°. Because unpolarized light is converted to circular polarized light, this configuration is called a circular polarizer. Referring to Figure 2.36, the Jones matrix of this stack is given by multiplying the Jones matrices for rotating the coordinates by 45°, retarding the phase of the vertical component by 45°, and rotating back the coordinates in this order. When a linearly polarized light $\begin{bmatrix} 0 \\ 1 \end{bmatrix}$ enters this configuration, the output is calculated as follows.

$$R\left(-\frac{\pi}{4}\right)\begin{bmatrix} 1 & 0 \\ 0 & -i \end{bmatrix} R\left(\frac{\pi}{4}\right)\begin{bmatrix} 0 \\ 1 \end{bmatrix} = \frac{1}{\sqrt{2}} \cdot \frac{1}{\sqrt{2}} \begin{bmatrix} 1+i \\ 1-i \end{bmatrix} \tag{2.24}$$

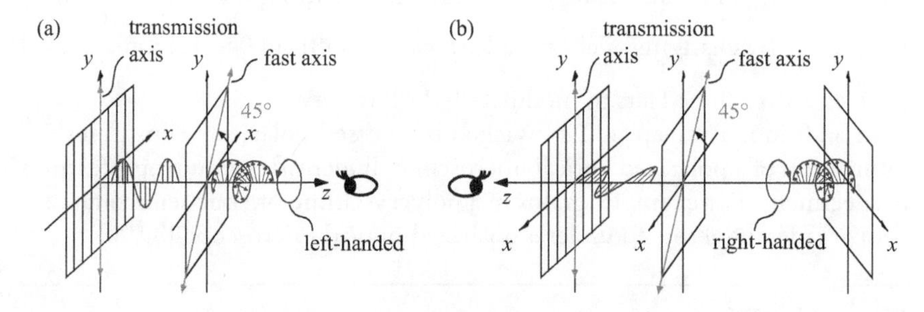

Figure 2.36 An example of polarization control. (a) A circular polarizer consists of a polarizer and a QWP. The angle between their optical axes is set to 45°. (b) Stacking a circular polarizer on a reflector prevents incoming light from being reflected.

Because the vertical component is delayed from the horizontal component by 45°, the expression above is equivalent to $\frac{1}{\sqrt{2}}\begin{bmatrix} 1 \\ i \end{bmatrix}$. This is a left-handed circular polarized light.

After reflection, the directions of the z and x axes are reversed. Now, the circular polarized light is right-handed, and it enters the rotated waveplate. Repeating a similar calculation, the output of the QWP is given by,

$$R\left(\frac{\pi}{4}\right)\begin{bmatrix} 1 & 0 \\ 0 & -i \end{bmatrix} R\left(-\frac{\pi}{4}\right)\frac{1}{\sqrt{2}}\begin{bmatrix} 1 \\ -i \end{bmatrix} = \frac{1}{\sqrt{2}}\begin{bmatrix} 1-i \\ 0 \end{bmatrix} \tag{2.25}$$

This is equivalent to $\begin{bmatrix} 1 \\ 0 \end{bmatrix}$. Hence, the light is linearly polarized along the horizontal axis and is absorbed by the polarizer.

If the retardation of the QWP deviates from 45°, elliptical polarization results. The polarizer transmits a part of it. Thus, the transmittance is determined by the retardation of the medium placed between the linear polarizer and the reflector. A reflective LCD utilizes this principle. Retardation of a liquid crystal layer is modulated by an external bias, and reflectance of ambient light is altered accordingly.

2.4.2 Liquid crystal devices

2.4.2.1 Liquid crystal phases

Liquid crystal (LC) materials have properties of both a liquid and a crystal at room temperature. In many cases, one can picture them as rod-like molecules. They organize themselves to form ordered structures due to

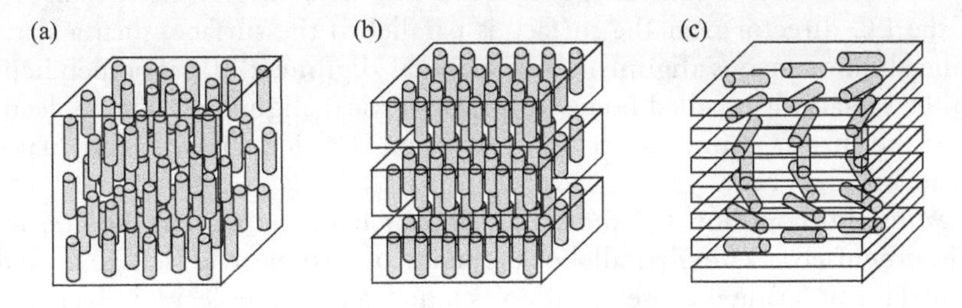

Figure 2.37 Interactions between molecules lead to ordered structures: (a) nematic phase, (b) smectic phase, and (c) cholesteric phase.

interactions with surrounding molecules. As a result, LC molecules exhibit some characteristic phases as illustrated in Figure 2.37. For example, in the nematic phase, molecules are aligned along one direction, but their centroids are randomly distributed as shown in Figure 2.37a. If they match each other, LC molecules form a layered structure as shown in Figure 2.37b. This phase is called smectic. If molecules in adjacent layers rotate continuously, they form a helix structure called a cholesteric phase as shown in Figure 2.37c. These orderings disappear at high temperatures and molecules are randomly oriented. This state is called isotropic phase.

2.4.2.2 Alignment

For practical applications, LC molecules need to be contained in a cell. By forming a thin polyimide layer on the surface of a container and making some textures on it, one can align LC molecules along them. This is schematically drawn in Figure 2.38. The thin layer formed for this purpose is called alignment layer. For example, nano-size grooves are created on a polyimide surface by rubbing it with a cloth. This alignment technique is called rubbing. It was widely used in the past despite its problems with dust particles which reduce the production yield of TFT arrays. A technique involving the formation of a photosensitive layer and oblique irradiation of polarized light is called photoalignment [91].

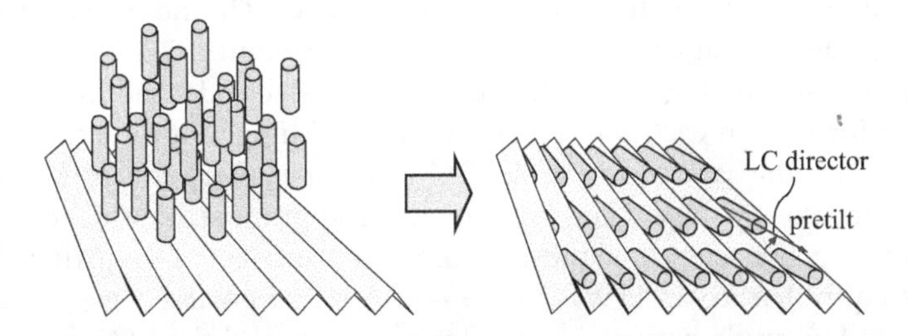

Figure 2.38 Molecule alignment at the surface of a container.

Average direction of the long axes of rod-like molecules is called LC director. If the LC director near the surface is parallel to the surface, such a state is called homogeneous alignment or horizontal alignment. If it is perpendicular to the surface, it is called homeotropic or vertical alignment. The molecules near the surface cannot move. This situation is called anchoring in analogy to anchoring a boat.

Although common LC molecules do not have permanent dipoles, they align themselves either parallel or perpendicular to electric field lines under bias. This ordering is caused by dipoles induced in LC molecules. Even when the field direction is reversed, LC molecules do not rotate. When driving an LCD, the polarity of the bias applied to the LC molecules in a pixel is periodically reversed to prevent impurity ions from accumulating at a certain region in the pixel. The LC molecules that align parallel to electric field lines are said to have positive dielectric anisotropy. Those that align perpendicularly are said to have negative dielectric anisotropy. For both materials, one needs to control which end of a rod-like molecule starts to tilt first. Otherwise, multiple domains would be created in a random fashion and light would leak through boundaries between them. Fortunately, rubbing a polymer film is a directional process and the LC director is not completely parallel to the surface. This angular deviation is called pretilt and LC molecules start to tilt from one end upon bias application.

Optical characteristics of LCDs are determined by how LC molecules are arranged in a pixel. By designing shapes and positions of electrodes and developing suitable alignment techniques, one can arrange LC molecules in an intended manner. Such arrangements developed so far include twisted nematic (TN), in-plane switching (IPS), multi-domain vertical alignment (MVA), etc. Because large-size displays are often watched from oblique directions, a wide viewing range is required. In this regard, IPS and MVA modes proposed in the 1990s played a critical role in competing with plasma display panel technology.

2.4.2.3 *Electrically controlled birefringence (ECB) mode*

Among various LC modes, the ECB mode is described first because its operation is analogous to the transmittance control with passive devices described above. Its applications for a light valve [92] and a passive-matrix LCD (i.e., no TFT at each pixel) [93] were reported in 1972.

As illustrated in Figure 2.39, a transparent electrode and an alignment layer are formed on each transparent substrate. Alignment directions at both substrates are the same. Hence, the LC director is parallel to the substrates throughout the LC layer, i.e., homogeneous alignment. Its thickness d and the birefringence $n_e - n_o$ of the LC material are selected such that the LC layer is equivalent to an HWP, i.e., $\Delta = \pi$. A linear polarizer is placed at the entrance. The angle between its transmission axis and the LC director is set to 45°. As expressed with the Jones vector below, linearly polarized light emerges. Note that the polarization plane is rotated by 90°.

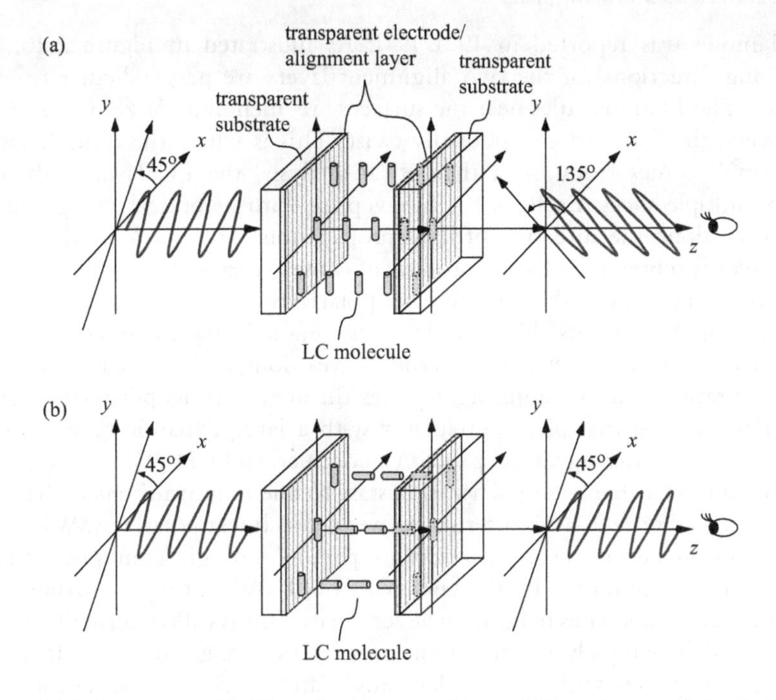

Figure 2.39 Operation of ECB mode: (a) at zero bias and (b) under a sufficiently high bias.

$$\begin{bmatrix} 1 & 0 \\ 0 & -1 \end{bmatrix} \frac{1}{\sqrt{2}} \begin{bmatrix} 1 \\ 1 \end{bmatrix} = \frac{1}{\sqrt{2}} \begin{bmatrix} 1 \\ -1 \end{bmatrix} \tag{2.26}$$

When a bias is applied, an electric field perpendicular to the substrates develops. If the electric field is high enough, LC molecules with positive dielectric anisotropy align themselves along the electric field, i.e., vertical alignment. The incident light passes through with its polarization state unaffected because the LC layer exhibits an ordinary refractive index irrespective of its polarization state.

Our eyes cannot distinguish circular and linearly polarized light. For modulating light intensity, one needs to add a linear polarizer for transmission control or a reflector for reflectance control.

Note that the argument above applies to the case of normal incidence. For oblique incidence, light propagates a longer distance in the birefringent medium and the phase retardation Δ increases. Because transmittance is a periodic function of Δ, it varies with the viewing angle periodically. Suppose that transmittance of a pixel in an LCD decreases with the viewing angle. Beyond a certain viewing angle, it starts to increase. This phenomenon is called grayscale inversion. Thus, the angular range that a viewer can watch a right image is limited. This poses a serious problem for a large-area display.

2.4.2.4 *Twisted nematic mode*

TN mode was reported in 1971 [94]. As illustrated in Figure 2.40, the rubbing directions for the two alignment layers are perpendicular to each other. The LC molecules near the surfaces are anchored. At zero bias, those between the two surfaces gradually twist. This is where the term "twisted nematic" comes from. In a theoretical analysis, the LC layer is divided into multiple slices. Each slice is a waveplate with its optical axis gradually rotating along the direction of light propagation. The Jones matrix of this LC layer is represented by the product of those for each slice. When linearly polarized light enters this structure, its polarization plane rotates 90°. When a bias is applied and its electric field is high enough, the LC molecules with positive dielectric anisotropy align themselves along it. This is a homeotropic configuration and incoming light passes through with its polarization state unaffected. A spatial phase modulator with a large dynamic range can be realized by modifying an LC panel in a commercial LCD [95].

To convert a change in polarization state to one in transmittance, the configuration (Figure 2.40) is inserted between two linear polarizers. When the transmission axes of these polarizers are parallel, no light comes out at zero bias. This configuration is called normally black. When they are orthogonal, transmittance becomes maximum at zero bias. This is called normally white.

Note that obliquely incident light propagates a longer distance. Its phase retardation varies with the incident angle and so does the transmittance.

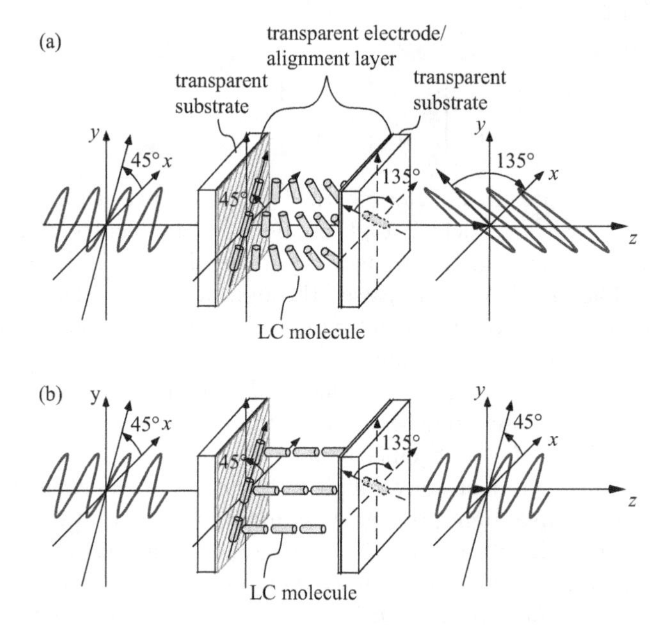

Figure 2.40 Twisted nematic mode: (a) at zero bias, and (b) under a sufficiently high bias.

Therefore, a viewing angle of a TN-mode LCD is limited. This poses a serious problem for a large-area LCD.

2.4.2.5 In-plane switching mode

IPS mode solves the problem of viewing angle by rotating LC molecules in a plane parallel to the substrates [96]. This is accomplished by a pair of electrodes fabricated on one substrate as illustrated in Figure 2.41. The alignment direction is along the y axis on both substrates so that LC molecules are aligned parallel to the y axis at zero bias. The birefringence $n_e - n_o$ and the thickness d are selected such that this LC layer is equivalent to a HWP, i.e., $\Delta = \pi$. It is sandwiched by crossed polarizers. The transmission axis of the polarizer on the entrance side is along the y axis. The light passes through the LC layer unaffected. The polarizer at the exit side absorbs it. When a bias is applied between the two electrodes, electric field lines are mostly parallel to the substrates. At a sufficiently high bias, LC molecules with positive dielectric anisotropy rotate in the xy plane and align themselves along the field lines. The LC layer in this case is equivalent to a HWP with its optical axis at 45° with the horizontal axis. Hence, the polarization plane is rotated by 90°, and the polarizer at the exit side transmits it.

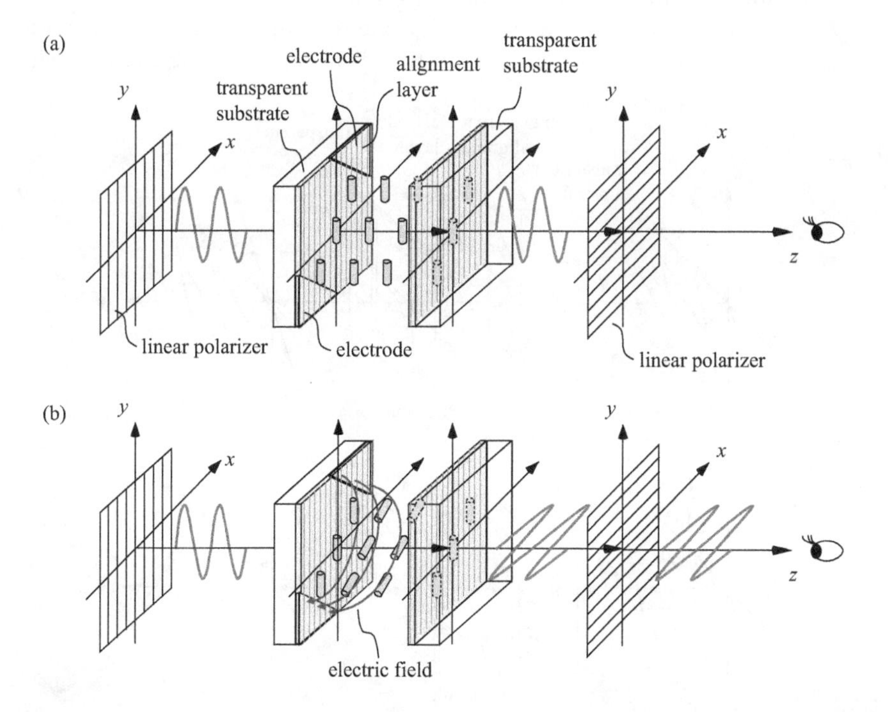

Figure 2.41 Operation of IPS mode: (a) at zero bias and (b) under a sufficiently high bias.

The magnitude of the electric field determines the rotation angle of the LC director, i.e., the optical axis of the equivalent HWP. Denoting this angle as θ, its transmittance is proportional to $\sin^2 2\theta$. Therefore, the transmittance can take full swing when θ is varied from 0° to 45°. It varies monotonically with the bias applied between the electrodes. Finally, it is interesting to note that the actual alignment direction is not exactly along the y axis. It is interpreted that the purpose of this slight deviation is to avoid defect formation. Namely, if it were exactly along the y axis, LC molecules have an equal chance to rotate in positive and negative directions, leading to formation of multiple domains with boundaries in between.

2.4.2.6 Multi-domain vertical alignment mode

Suppose that light is incident on an ECB mode LC cell. The direction of incidence is specified by polar and azimuthal angles. When LC molecules are vertically aligned, phase retardation by the LC layer depends on the polar angle only. When they are tilted, it depends on the azimuthal angle as well. The dependency on the azimuthal angle can be lessened by averaging, i.e., dividing the incident area into two or four regions with compensating characteristics and adding the light coming out from each region. This configuration is called multi-domain vertical alignment (MVA).

An example of a four-domain configuration using an LC material with negative dielectric anisotropy is illustrated in Figure 2.42. Alignment

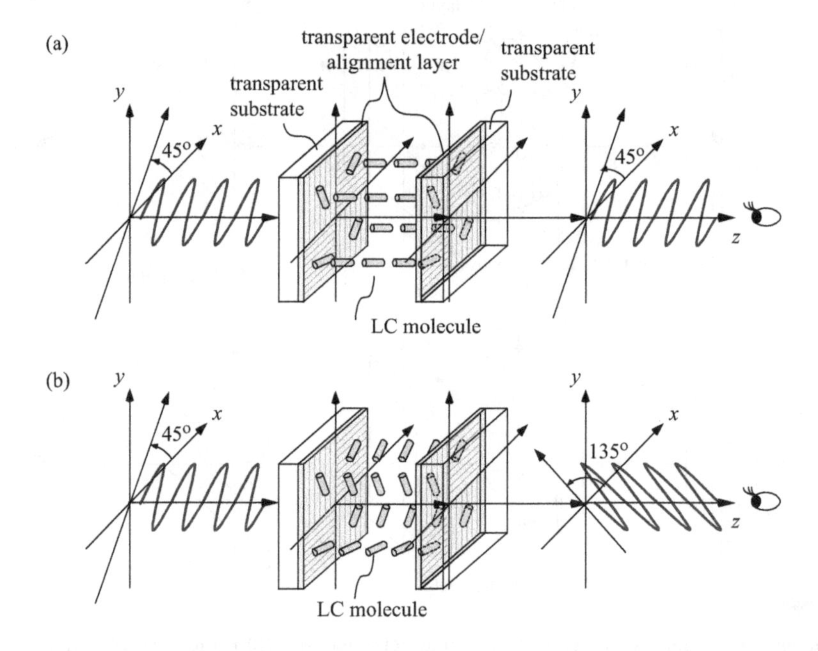

Figure 2.42 Operation of MVA mode: (a) at zero bias and (b) under a sufficiently high bias. Note that the LC material is assumed to have negative dielectric anisotropy.

directions in the first and third quadrants are along the x axis while those in the second and fourth quadrants are along the y axis. Pretilt angles in the first and third quadrants have opposite signs. So do those in the second and fourth quadrants. Thus, LC directors in the four quadrants under a sufficiently high bias have rotational symmetry around the z axis.

At zero bias, LC molecules are vertically aligned as shown in Figure 2.42a. Because this is equivalent to an isotropic medium, light passes through with its polarization state unchanged. When a bias is applied and the electric field is high enough, LC molecules are aligned perpendicular to the field lines as shown in Figure 2.42b. Note that this is a material with negative dielectric anisotropy. Each quadrant is an HWP. The angle between its optical axis and the polarization of incident light is 45°. Linearly polarized light emerges from each of the four quadrants. Each polarization plane is rotated by 90°. Normally black operation is realized by placing a linear polarizer at the exit side and by setting the angle between its optical axis and the x axis to 135°. Because of the rotational symmetry, dependency of polarization state on azimuthal angle of incident direction is compensated by adding the light emerging from the diagonal quadrant pairs.

For practical LCD applications, it is desired to create multiple LC domains in each pixel without rubbing process. Fujitsu reported an approach utilizing protruded regions in 1998 [97]. Protruded regions were fabricated on both substrates before forming alignment layers. The slopes set the directions for LC molecules to tilt. In 2010, Sharp reported another rubbing-less technique. It was based on photoalignment. Irradiation of polarized ultraviolet light on a polyimide film containing photosensitive materials resulted in a pretilt angle of 1° from the plane normal [98].

2.4.3 MEMS devices

One can fabricate an overhanging structure such as a cantilever with MEMS technology and make its surface reflective. Such a mirror can be oscillated by applying electrostatic force, electromagnetic force, piezoelectric force, etc. Turning off the force brings back the mirror to its original position. Therefore, incoming light is deflected in different directions. If the mirror changes its position continuously, it can scan a laser beam. If only two positions are exploited, the mirror works as an on-off optical switch. Three examples of MEMS-based devices are described below.

2.4.3.1 DMDs

Texas Instruments has a long history of developing DMDs [99]. Initially, it was an analog device called "deformable membrane device." One-dimensional arrays were studied for printer applications. The concept of bistable "digital" operation was conceived in 1987. This device is a two-dimensional array of oscillating mirrors. An early version of its pixel structure is illustrated in Figure 2.43. Two torsion hinges support a mirror. Two addressing electrodes

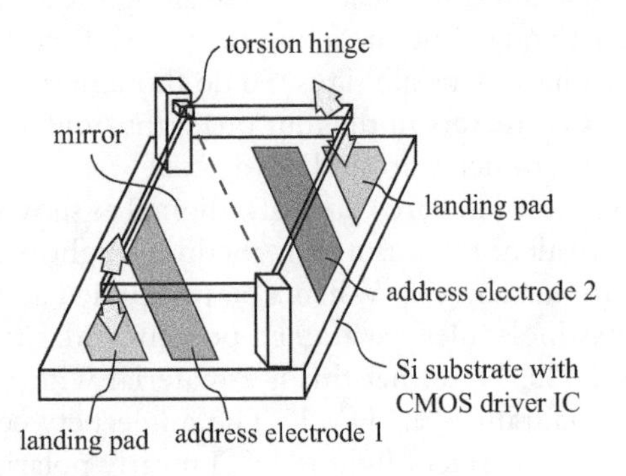

Figure 2.43 Basic structure of a pixel in an early DMD.

and two landing pads are provided underneath the mirror. Driver circuits are fabricated on a Si substrate by standard CMOS processes prior to mirror fabrication by MEMS technology. When a clock pulse with an opposite phase is supplied to each addressing electrode, one of them attracts the mirror and the other repels it. As a result, one end of the mirror lands on one of the landing pads. In this way, incoming light is directed in two possible stable directions. Grayscale images are displayed by pulse width modulation, i.e., controlling the time duration for the mirror at each position.

Aiming at higher reflectance and reliability, the design of a DMD has advanced. One notable example is called "hidden hinges": a separate mirror is attached to an oscillating plate. By minimizing the space between adjacent mirrors, the area ratio of a mirror and a pixel increases. It is interesting to note that a similar "umbrella structure" is used for increasing the fill factor of microbolometer-based IR image sensors (see Section 4.5.2).

2.4.3.2 Grating light valves

An electrically switchable phase grating is fabricated by dividing a suspended mirror into periodic stripes and providing a common electrode beneath them [100]. Each mirror is supported at both ends. At zero bias, each mirror reflects incident light. When a bias is applied between the stripe and the common electrode, it is deformed with its central region relatively flat. The displacement of the central region of the mirror is designed to be equal to a quarter of the wavelength of incident light. When alternate stripes are biased as illustrated in Figure 2.44, the structure becomes equivalent to a reflection phase grating.

Note that diffraction characteristics are explicitly dependent on the wavelength of light. Hence, this device is suitable for controlling a monochromatic laser beam. One-dimensional arrays are used together with beam-scanning optics for laser display applications (see Section 5.4.2). Because the stripes must be longer than the relatively flat central regions, it is not practical to make two-dimensional arrays with this technology.

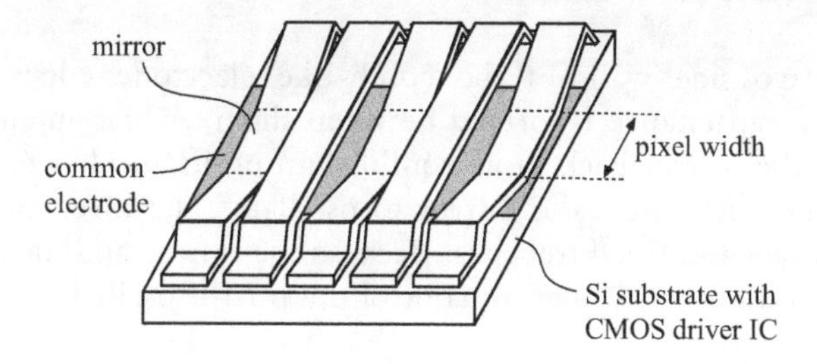

Figure 2.44 Basic structure of a pixel in a grating light valve (GLV).

2.4.3.3 Laser beam scanners

A small two-dimensional scanner is built into a mobile laser projector [101]. Despite its drawback of speckle noise, it attracted much attention since the mid-2010s when green semiconductor laser diodes had become available. There are other applications for scanning laser beams such as printing, retinal scanning displays, and light detection and ranging (LiDAR), where MEMS-based laser scanners are suited for miniaturization of systems. There are three types of forces to rotate mirrors: Coulomb force, Lorentz force, and piezoelectric force [102].

For example, a two-dimensional scanner based on electrostatic actuation is illustrated in Figure 2.45 [103]. The device is fabricated on a silicon-on-insulator (SOI) wafer by a CMOS-compatible process. The top layer is a 30 μm-thick heavily doped Si. It serves as an electrode. The mirror at the center is supported by a pair of torsion hinges. It is fixed to the movable frame by the buried oxide layer of the SOI wafer. The movable frame is connected

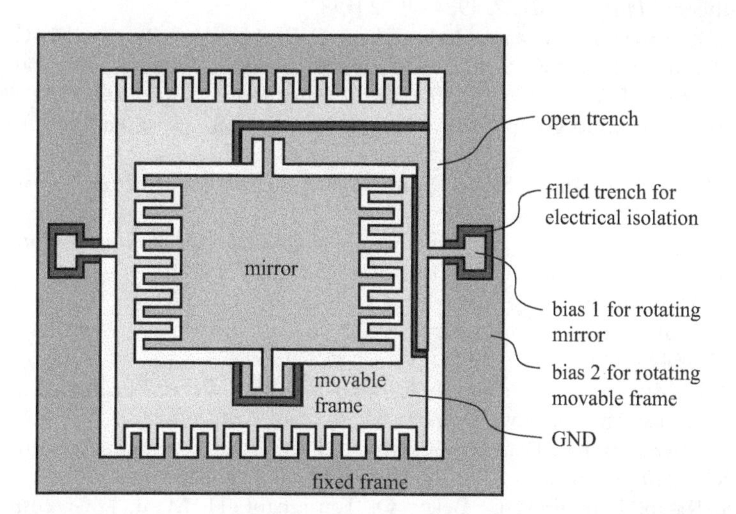

Figure 2.45 Top view of a two-dimensional MEMS scanner utilizing electrostatic force to rotate a mirror around horizontal and vertical axes independently.

to the outer frame in the same manner. This implementation is called gimbal mount.

Because two side walls of the comb-like electrode edges are facing each other, a capacitance is formed between them. A bias applied between them forces the mirror back to its equilibrium position. Hence, a bias with alternating polarity forces the mirror to oscillate. The three electrodes are isolated by open and filled trenches. Because the mirror and the fixed frame can be biased differently, one can control the mirror oscillation around two axes independently.

References

1. S. M. Sze, *Physics of Semiconductor Devices*, 2nd edition, John Wiley & Sons, 1981.
2. S. O. Kasap, *Principles of Electronic Materials and Devices*, 2nd edition, McGraw Hill, 2002.
3. G. F. Knoll, *Radiation Detection and Measurement*, John Wiley & Sons, 1979.
4. E. Fred Schubert, *Light-Emitting Diodes*, 2nd edition, Cambridge University Press, 2006.
5. C. C. Hu, *Modern Semiconductor Devices for Integrated Circuits*, Prentice Hall, 2009.
6. E. Hecht, *Optics*, 4th edition, Addison Wesley, 2002.
7. D. E. Carlson and C. R. Wronski, "Amorphous silicon solar cell," *Appl. Phys. Lett.* 28, 671–672 (1976).
8. R. S. Crandall, "Transport in hydrogenated amorphous silicon p-i-n solar cells," *J. Appl. Phys.* 53, 3350–3352 (1982).
9. S. Ramo, "Currents induced by electron motion," *Proc. IRE.* 27(9), 584–585 (1939).
10. R. A. Street, "Measurements of depletion layers in hydrogenated amorphous silicon," *Phys. Rev. B* 27, 4924–4932 (1983).
11. Y. K. Fang, S-B. Hwang, K-H. Chen, C-R. Liu, M-J. Tsai, and L-C. Kuo, "An amorphous SiC/Si heterojunction p-i-n diode for low-noise and high-sensitivity UV detector," *IEEE Trans. Electron Devices* 39(2), 292–296 (1992).
12. http://www.nist.gov/pml/data/xray_gammaray.cfm (accessed on Aug. 5, 2022.)
13. F. S. Goulding and Y. Stone, "Semiconductor radiation detectors," *Science*, 170, 280–289 (1970).
14. C. A. Klein, "Bandgap dependence and related features of radiation ionization energies in semiconductors," *J. Appl. Phys.* 39, 2029–2038 (1968).
15. R. Devanathan, L. R. Corrales, F. Gao, and W. J. Weber, "Signal variance in gamma-ray detectors – A review," *Nucl. Instrum. Methods Phys. Res. A* 565, 637–649 (2006).
16. B. D. Milbrath, A. J. Peurrung, M. Bliss, and W. J. Weber, "Radiation detector materials: An overview," *J. Mater.* 23(10), 2561–2581 (2008).
17. A. Owens and A. Peacock, "Compound semiconductor radiation detectors," *Nucl. Instrum. Methods Phys. Res. A* 531(2), 18–37 (2004).
18. S. Kasap, J. B. Frey, G. Belev, O. Tousignant, H. Mani, J. Greenspan, L. Laperriere, O. Bubon, A. Reznik, G. De Crescenzo, K. S. Karim, and

J. A. Rowlands, "Amorphous and polycrystalline photoconductors for direct conversion flat panel x-ray image sensors," *Sensors* 11, 5112–5157 (2011).

19. T. J. Hajagos, C. Liu, N. J. Cherepy, and Q. Pei, "High-Z sensitized plastic scintillators: A review," *Adv. Mater.* 30, 1706956 (2018).
20. D. R. Schaart, "Physics and technology of time-of-flight PET detectors," *Phys. Med. Biol.* 66, 09TR01 (2021).
21. T. E. Peterson and L. R. Furenlid, "SPECT detectors: The Anger Camera and beyond," *Phys. Med. Biol.* 56(17), R145–R182 (2011).
22. There is an interesting story about its discovery by Emilio Segre and Glenn Seaborg. See W. C. Eckelman, "Unparalleled contribution of Technetium-99m to medicine over 5 decades," *J. Am. Coll. Cardiol. Img.* 2(3), 364–368 (2009).
23. P. Buzhan, B. Dolgoshein, L. Filatov, A. Ilyin, V. Kantzerov, V. Kaplin, A. Karakash, F. Kayumov, S. Klemin, E. Popova, and S. Smirnov, "Silicon photomultiplier and its possible applications," *Nucl. Instrum. Methods Phys. Res. A* 504, 48–52 (2003).
24. S. Gundacker and A. Heering, "The silicon photomultiplier: Fundamentals and applications of a modern solid-state photon detector," *Phys. Med. Biol.* 65, 17TR01 (2020).
25. I. Fujieda, G. Cho, J. Drewery, T. Gee, T. Jing, S. N. Kaplan, V. Perez-Mendez, and D. Wildermuth, "X-ray and charged particle detection with CsI(Tl) layer coupled to a Si:H photodiode layers," *IEEE Trans. Nucl. Sci.* 38(2), 255–262 (1991).
26. V. V. Nagarkar, J. S. Gordon, S. Vasile, P. Gothoskar, and F. Hopkins, "High resolution X-ray sensor for non-destructive evaluation," *IEEE Trans. Nucl. Sci.* 3(3), 1559–1563 (1996).
27. J. Kaniewski and J. Piotrowski. "InGaAs for infrared photodetectors. Physics and technology," *Opto-electron. Rev.* 12(1), 139–148 (2004).
28. V. Pejovi, E. Georgitzikis, J. Lee, I. Lieberman, D. Cheyns, P. Heremans, and P. E. Malinowski, "Infrared colloidal quantum dot image sensors," *IEEE Trans. Electron Devices* 69(6), 2840–2850 (2022).
29. A. Rogalski, "Progress in focal plane array technologies," *Prog. Quantum. Electron.* 36(2–3), 342–473 (2012).
30. A. Rogalski, "HgCdTe infrared detector material: History, status and outlook," *Rep. Prog. Phys.* 68, 2267–2336 (2005).
31. D. B. Reynolds, D. H. Seib, S. B. Stetson, T. Herter, N. Rowlands, and J. Schoenwald, "Blocked impurity band hybrid infrared focal plane arrays for astronomy," *IEEE Trans. Nucl. Sci.* 36(1), 857–862 (1989).
32. J. E. Huffman, A. G. Crouse, B. L. Halleck, T. V. Downes, and T. L. Herter, "Si:Sb blocked impurity band detectors for infrared astronomy," *J. Appl. Phys.* 72, 273–275 (1992).
33. P. J. Love, A. W. Hoffman, N. A. Lum, K. J. Ando, J. Rosbeck, W. D. Ritchie, N. J. Therrien, R. S. Holcombe, and E. Corrales, "1024×1024 Si:As IBC detector arrays for JWST MIRI," *Proc. SPIE* 5902, 590209 (2005).
34. B. F. Levine, "Quantum-well infrared photodetectors," *J. Appl. Phys.* 74, R1R81 (1993).
35. A. Rogalski, "Quantum well photoconductors in infrared detector technology," *J. Appl. Phys.* 93, 4355–4391 (2003).
36. K. C. Liddiard, "Thin-film resistance bolometer IR detectors", *Infrared Phys.* 24, 57–64 (1984).

37. F. Niklaus, C. Vieider, and H. Jakobsen, "MEMS-based uncooled infrared bolometer arrays: a review," *Proc. SPIE 6836, MEMS/MOEMS Technologies and Applications III*, 2008, p. 68360D.
38. L. F. Weber, "History of the plasma display panel," *IEEE Trans. Plasma Sci.* 34(2), 268–278 (2006).
39. R. Dressler, "The PDF Chromatron-a single or multi-gun tri-color Cathode-Ray Tube," *Proceed. IRE* 41(7), 851–858 (1953).
40. A. A. Talin, K. A. Dean, and J. E. Jaskie, "Field emission displays: A critical review," *Solid State Electron.* 45(6), 963–976 (2001).
41. R. D. Dupuis and M. R. Krames, "History, development, and applications of high-brightness visible Light-Emitting Diodes," *J. Light* 26(9), 1154–1171 (2008).
42. A. I. Zhmakin, "Enhancement of light extraction from light emitting diodes," *Phys. Rep.* 498(4–5), 189–241 (2011).
43. J. Cho, J. H. Park, J. K. Kim, and E. Fred Schubert, "White light-emitting diodes: History, progress, and future," *Laser Photonics Rev.* 11(2), 1600147 (2017).
44. H. S. El-Ghoroury, Y. Nakajima, M. Yeh, E. Liang, C-L. Chuang, and J. C. Chen, "Color temperature tunable white light based on monolithic color-tunable light emitting diodes," *Opt. Express* 28, 1206–1215 (2020).
45. B. Geffroy, P. le Roy, and C. Prat, "Organic light-emitting diode (OLED) technology: Materials, devices and display technologies," *Polym. Int.* 55, 572–582 (2006).
46. C. W. Tang and S. A. VanSlyke, "Organic electroluminescent diodes," *Appl. Phys. Lett.* 51, 913 (1987).
47. A. P. Ghosh, L. J. Gerenser, C. M. Jarman, and J. E. Fornalik, "Thin-film encapsulation of organic light-emitting devices," *Appl. Phys. Lett.* 86, 223503 (2005).
48. S. Lee, J-H. Han, S-H. Lee, and J-S. Park, "Review of organic/inorganic thin film encapsulation by atomic layer deposition for a flexible OLED display," *JOM* 71, 197–211 (2019).
49. Y. Takafuji, Y. Fukushima, K. Tomiyasu, M. Takei, Y. Ogawa, K. Tada, S. Matsumoto, H. Kobayashi, Y. Watanabe, E. Kobayashi, S. R. Droes, A. T. Voutsas, and J. Hartzell, "Integration of single crystal Si TFTs and circuits on a large glass substrate," *Proc. IEEE Int. Electron Devices Meet. (IEDM)*, 1–4 (2009).
50. G. Fortunato, A. Pecora, and L. Maiolo, "Polysilicon thin-film transistors on polymer substrates," *Mater. Sci. Semicond. Process* 15(6), 627–641 (2012).
51. S. Takagi, A. Toriumi, M. Iwase, and H. Tango, "On the universality of inversion layer mobility in Si MOSFET's: Part I-effects of substrate impurity concentration," *IEEE Trans. Electron Devices* 41(12), 2357–2362 (1994).
52. C. Hu, S. C. Tam, F-C. Hsu, P-K. Ko, T-Y. Chan, and K. W. Terrill, "Hot-electron-induced MOSFET degradation-model, monitor, and improvement," *IEEE J. Solid-State Circuits* 20(1), 295–305 (1985).
53. S. Ogura, P. J. Tsang, W. W. Walker, D. L. Critchlow, and J. F. Shepard, "Design and characteristics of the lightly doped drain-source (LDD) insulated gate field-effect transistor," *IEEE J. Solid-State Circuits* 15(4), 424–432 (1980).
54. M. V. Fischetti, D. A. Neumayer, and E. A. Cartier, "Effective electron mobility in Si inversion layers in metal-oxide-semiconductor systems with a high-κ insulator: The role of remote phonon scattering," *J. Appl. Phys.* 90, 4587–4608 (2001).

55. R. People, "Physics and applications of GexSi1-x/Si strained-layer heterostructures," *IEEE J. Quantum Electron.* 22(9), 1696–1710 (1986).
56. J. Welser, J. L. Hoyt, and J. F. Gibbons, "Electron mobility enhancement in strained-Si n-type metal-oxide-semiconductor field-effect transistors," *IEEE Electron Device Lett.* 15(3), 100–102 (1994).
57. P. G. LeComber, W. E. Spear, and A. Ghaith, "Amorphous silicon field-effect device and possible application," *Electron. Lett.* 15, 179–181 (1979).
58. A. J. Snell, K. D. Mackenzie, W. E. Spear, and P. G. LeComber, "Application of amorphous silicon field effect transistors in addressable liquid crystal display panels," *Appl. Phys.* 24, 357–362 (1981).
59. M. J. Powell, "The physics of amorphous-silicon thin-film transistors," *IEEE Electron Device Lett.* 36(12), 2753–2763 (1989).
60. J. Souk, S. Morozumi, F-C. Luo, and I. Bita, *Flat Panel Display Manufacturing*, Wiley, 2018, p. 8.
61. T. Sameshima, S. Usui, and M. Sekiya, "XeCl Excimer laser annealing used in the fabrication of poly-Si TFT's," *IEEE Electron Device Lett.* 7(5), 276–278 (1986).
62. K. Sera, F. Okumura, H. Uchida, S. Itoh, S. Kaneko, and K. Hotta, "High-performance TFTs fabricated by XeCl excimer laser annealing of hydrogenated amorphous-silicon film," *IEEE Trans. Electron Devices* 36(12), 2868–2872 (1989).
63. G. Rajeswaran, M. Itoh, M. Boroson, S. Barry, T. K. Hatwar, K. B. Kahen, K. Yoneda, R. Yokoyama, T. Yamada, N. Komiya, H. Kanno, and H. Takahashi, "Active matrix low temperature poly-Si TFT / OLED full color displays: Development status," *SID Symp. Dig. Tech. Pap.* 31, 974–977 (2000).
64. K. Nomura, H. Ohta, A. Takagi, T. Kamiya, M. Hirano, and H. Hosono, "Room-temperature fabrication of transparent flexible thin-film transistors using amorphous oxide semiconductors," *Nature* 432, 488–492 (2004).
65. E. Fortunato, P. Barquinha, and R. Martins, "Oxide semiconductor thin-film transistors: A review of recent advances," *Adv. Mater.* 24, 2945–2986 (2012).
66. J. F. Wager, "Oxide TFTs: A progress report," *Inform. Disp.* 32(1), 1–16 (2016).
67. J. F. Wager, "TFT technology: Advancements and opportunities for improvement," *Inform. Disp.* 36(1), 9–13 (2020).
68. R. A. Street, "Thin-film transistors," *Adv. Mater.* 21, 2007–2022 (2009).
69. R. A. Street, T. N. Ng, D. E. Schwartz, G. L. Whiting, J. P. Lu, R. D. Bringans, and J. Veres, "From printed transistors to printed smart systems," *Proc. IEEE* 103(4), 607–618 (2015).
70. H. Minemawari, T. Yamada, H. Matsui, J. Tsutsumi, S. Haas, R. Chiba, R. Kumai, and T. Hasegawa, "Inkjet printing of single-crystal films," *Nature* 475, 364–367 (2011).
71. J. E. Northrup, "Two-dimensional deformation potential model of mobility in small molecule organic semiconductors," *Appl. Phys. Lett.* 99, 062111 (2011).
72. P. Lin, and F. Yan, "Organic thin-film transistors for chemical and biological sensing," *Adv. Mater.* 24, 34–51 (2012).
73. M. Jung, J. Kim, J. Noh, N. Lim, C. Lim, G. Lee, J. Kim, H. Kang, K. Jung, A. D. Leonard, J. M. Tour, and G. Cho, "All-printed and roll-to-roll-printable 13.56-MHz-operated 1-bit RF tag on plastic foils," *IEEE Trans. Electron Devices* 57(3), 571–580 (2010).

74. G. Lavareda, C. Nunes de Carvalho, A. Amaral, E. Fortunato, A. R. Ramos, and M. F. da Silva, "Dependence of TFT performance on the dielectric characteristics," *Thin Solid Films* 427(1–2), 71–76 (2003).
75. K. S. Karim, A. Nathan, M. Hack, and W. I. Milne, "Drain-bias dependence of threshold voltage stability of amorphous silicon TFTs," *IEEE Electron Device Lett.* 25(4), 188–190 (2004).
76. Y. Oana, "Current and future technology of low-temperature poly-Si TFT-LCDs," *J. Soc. Inf. Display* 9, 169–172 (2001).
77. Y. Morimoto, Y. Jinno, K. Hirai, H. Ogata, T. Yamada, and K. Yoneda, "Influence of the grain boundaries and intragrain defects on the performance of poly-Si thin film transistors," *J. Electrochem. Soc.* 144, 2495–2501 (1997).
78. A. Hara, F. Takeuchi, M. Takei, K. Suga, K. Yoshino, M. Chida, Y. Sano, and N. Sasaki, "High-performance polycrystalline silicon thin film transistors on non-alkali glass produced using continuous wave laser lateral crystallization," *Jpn. J. Appl. Phys.* 41, L311 (2002).
79. T. Noguchi, Y. Chen, T. Miyahira, J. de Dieu Mugiraneza, Y. Ogino, Y. Iida, E. Sahota, and M. Terao, "Advanced micro-polycrystalline silicon films formed by blue-multi-laser-diode annealing.," *Jpn. J. Appl. Phys.* 49, 03CA10 (2010).
80. S. Lee, Y. Do, D. Kim, and J. Jang, "Extremely foldable LTPS TFT backplane using blue laser annealing for low-cost manufacturing of rollable and foldable AMOLED display," *J. Soc. Inf. Display.* 29, 382–389 (2021).
81. N. M. Johnson, D. K. Biegelsen, H. C. Tuan, M. D. Moyer, and L. E. Fennell, "Single-crystal silicon transistors in laser-crystallized thin films on bulk glass," *IEEE Electron Device Lett.* 3(12), 369–372 (1982).
82. J. C. Park, S. W. Kim, S. Il Kim, H. Yin, J. H. Hur, S. H. Jeon, S. H. Park, I. H. Song, Y. S. Park, U. I. Chung, M. K. Ryu, S. Lee, S. Kim, Y. Jeon, D. M. Kim, D. H. Kim, K-W. Kwon, and C. J. Kim, "High performance amorphous oxide thin film transistors with self-aligned top-gate structure," *Proc. IEEE Int. Electron Devices Meet. (IEDM)*, 191–194 (2009).
83. C. Chen, B-R. Yang, C. Liu, X-Y. Zhou, Y-J. Hsu, Y-C. Wu, J-S. lm, P-Y. Lu, M. Wong, H-S. Kwok, and H-P. D. Shieh, "Integrating poly-silicon and InGaZnO thin-film transistors for CMOS inverters," *IEEE Trans. Electron Devices* 64(9), 3668–3671 (2017).
84. D. Y. Jeong, Y. Chang, W. G. Yoon, Y. Do, and J. Jang, "Low-temperature polysilicon oxide thin-film transistors with coplanar structure using six photomask steps demonstrating high inverter gain of $264\,V\,V^{-1}$," *Adv. Eng. Mater.* 22, 1901497 (2020).
85. K. Nomura, "Recent progress of oxide-TFT-based inverter technology," *J. Inf. Disp.* 22(4), 211–229 (2021).
86. R. A. Street, T. N. Ng, R. A. Lujan, I. Son, M. Smith, S. Kim, T. Lee, Y. Moon, and S. Cho, "Sol-gel solution-deposited InGaZnO thin film transistors," *ACS Appl. Mater. Interfaces* 6(6), 4428–4437 (2014).
87. Y. Ukai, T. Ohyama, L. Fennell, Y. Kato, M. Paukshto, P. Smith, O. Yamashita, and S. Nakanishi, "Current status and future prospect of in-cell-polarizer technology," *J. Soc. Inf. Display* 13(1), 17–24 (2005).
88. H. Chen, G. Tan, and S-T. Wu, "Ambient contrast ratio of LCDs and OLED displays," *Opt. Express* 25(26), 33643–33656 (2017).
89. I. Fujieda, N. Iizuka, and Y. Onishi, "Directional solidification of C8-BTBT films induced by temperature gradients and its application for transistors," *Proc. SPIE* 9360, 936012 (2015).

90. N. Iizuka, T. Zanka, Y. Onishi, and I. Fujieda, "Growth directions of C8-BTBT thin films during drop-casting," *Proc. SPIE* 9745, 97451J (2016).
91. K. Ichimura, "Photoalignment of liquid-crystal systems," *Chem. Rev.* 100(5), 1847–1874 (2000).
92. R. A. Soref and M. J. Rafuse, "Electrically controlled birefringence of thin nematic films," *J. Appl. Phys.* 43, 2029–2037 (1972).
93. M. Hareng, G. Assouline, and E. Leiba, "Liquid crystal matrix display by electrically controlled birefringence," *Proc. IEEE* 60(7), 913–914 (1972).
94. M. Schadt and W. Helfrich, "Voltage-dependent optical activity of a twisted nematic liquid crystal," *Appl. Phys. Lett.* 18, 127 (1971).
95. K. Lu and B. E. A. Saleh, "Theory and design of the liquid crystal TV as an optical spatial phase modulator," *Opt. Eng.* 29(3), 240–245 (1990).
96. M. Oh-e and K. Kondo, "Electro-optical characteristics and switching behavior of the in-plane switching mode," *Appl. Phys. Lett.* 67, 3895 (1995).
97. A. Takeda, S. Kataoka, T. Sasaki, H. Chida, H. Tsuda, K. Ohmuro, T. Sasabayashi, Y. Koike, and K. Okamoto, "A super-high image quality multi-domain vertical alignment LCD by new rubbing-less technology," *SID Symp. Dig. Tech. Pap.* 29, 1077–1080 (1998).
98. K. Miyachi, K. Kobayashi, Y. Yamada, and S. Mizushima, "The world's first photo alignment LCD technology applied to Generation Ten factory," *SID Symp. Dig. Tech. Pap.* 41, 579–582 (2010).
99. L. J. Hornbeck, "From cathode rays to digital micromirrors: A history of electronic projection display technology," *TI Tech. J.* 15(3), 7–46 (1998).
100. O. Solgaard, F. Sandejas, and D. Bloom, "Deformable grating optical modulator," *Opt. Lett.* 17(9), 688–690 (1992).
101. W. O. Davis, R. Sprague, and J. Miller, "MEMS-based pico projector display," *2008 IEEE/LEOS International Conference on Optical MEMs and Nanophotonics*, 2008, pp. 31–32.
102. S. T. S. Holmström, U. Baran, and H. Urey, "MEMS laser scanners: A review," *J. Microelectromech. Syst.* 23(2), 259–275 (2014).
103. H. Schenk, P. Dürr, D. Kunze, H. Lakner, and H. Kück, "A resonantly excited 2D-micro-scanning-mirror with large deflection," *Sens. Actuators A, Phys.* 89(1), 104–111 (2001).

3 Thin-film semiconductors

Photodiodes, LEDs, and thin-film transistors (TFTs) are building blocks of LCDs, OLED displays, X-ray imagers, etc. For understanding and improving the characteristics of these devices, knowledge of thin-film semiconductor materials is required. These materials are either amorphous or polycrystalline and are fabricated on large-area substrates. Standard models on crystalline semiconductors are modified to explain properties of these materials. Furthermore, how we fabricate these materials on a large-area substrate with adequate uniformity at high speed is critically important for practical applications. Incidentally, energy crisis in the past prompted intensive studies on amorphous and polycrystalline semiconductors for solar cell applications. It's interesting to note that the vast knowledge accumulated by these studies has benefited display and other industries utilizing thin-film semiconductors.

3.1 Hydrogenated amorphous silicon

3.1.1 Film growth

Amorphous silicon (a-Si) prepared by glow discharge decomposition of silane is an electronic material [1]. As illustrated in Figure 3.1, silane (SiH_4) and hydrogen (H_2) gases are introduced in a vacuum chamber. They are

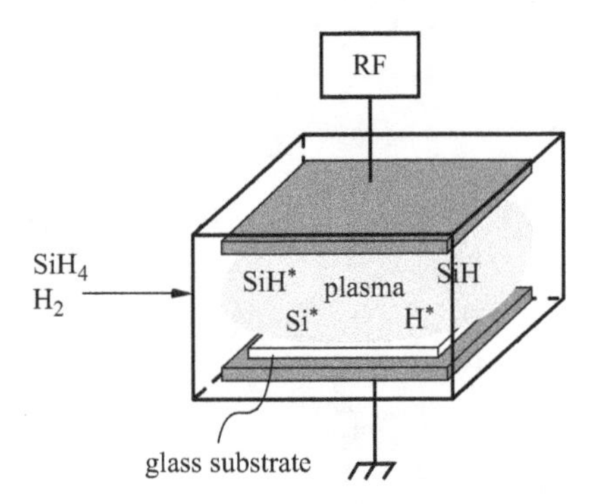

Figure 3.1 Preparation of an a-Si thin film by PECVD.

DOI: 10.1201/9781003364979-3

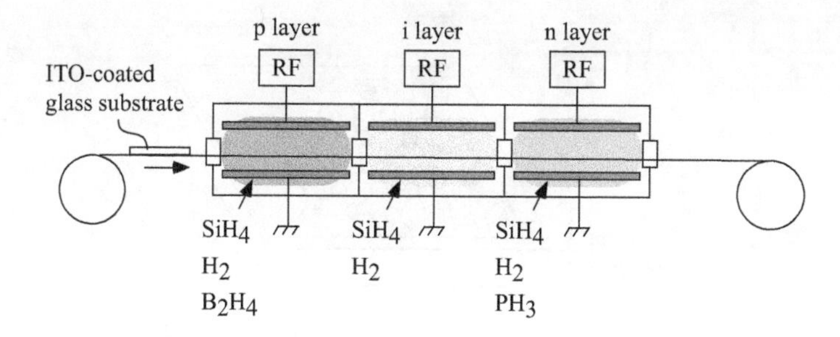

Figure 3.2 Sequential deposition of p-type, intrinsic, and n-type a-Si layers without exposure to air.

decomposed in a plasma generated by a radio frequency (RF) power source. Various radicals and ions called precursors exist in the plasma. Assisted by the energy from a heated substrate, they move around and settle on it. A rigid network of Si atoms is formed. This film formation process is called plasma-enhanced chemical vapor deposition (PECVD).

Mixing phosphine (PH_3) and diborane (B_2H_6) in source gases results in an n-type and p-type a-Si film, respectively. For fabricating solar cells and photodiodes, a transparent conductive oxide, typically indium tin oxide (ITO), is formed on a glass substrate. Three types of a-Si layers are deposited on it sequentially as illustrated in Figure 3.2, resulting in a p-i-n configuration.

3.1.2 Electronic properties

Conceptional drawing in Figure 3.3 shows how atoms are arranged in the film deposited by PECVD. There are four bonds for a Si atom, which form a network with neighboring atoms. However, interatomic distance is not constant in an amorphous material. Occasionally, there is an unsatisfied bond called dangling bond. During the deposition process, hydrogen atoms bond to dangling bonds and make them electronically inactive. Therefore, the number of dangling bonds decreases by this terminating process. Unterminated dangling bonds trap electrons. They become negatively charged and represent fixed space charges. To emphasize the effect of hydrogen atoms, the resultant material is often called hydrogenated a-Si and denoted as a-Si:H.

To understand experimental data, the picture in Figure 3.4 is accepted for the density of states (DOS) and carrier transport in amorphous semiconductors [2]. Dangling bonds create the states in the middle of the band gap. Lack of long-range order in an amorphous material results in the states near the band edges. They are called mid-gap states and tail states, respectively. Electron transport is explained by multiple trapping model. Electrons in the conduction band are captured by the tail states near the band edge. The trapped

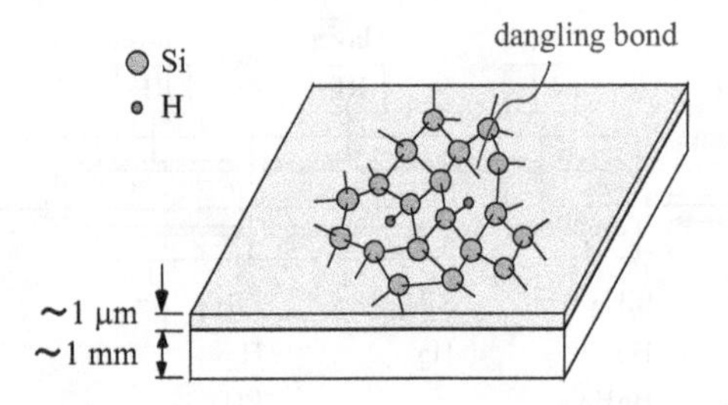

Figure 3.3 Conceptional drawing of an a-Si network to show how atoms bond to each other.

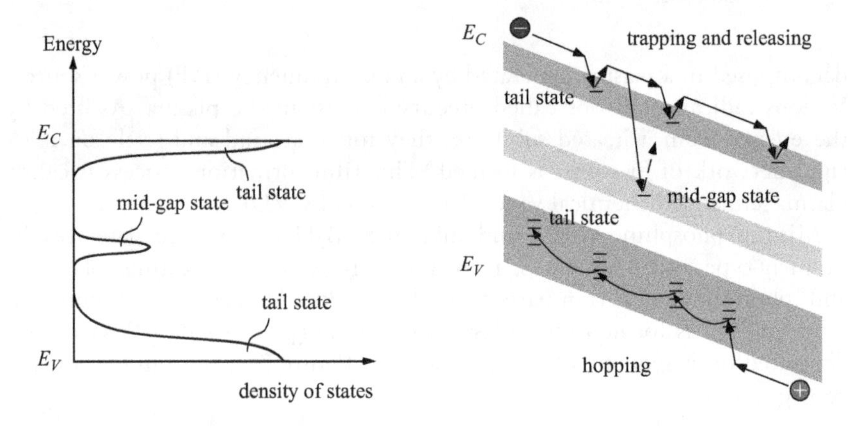

Figure 3.4 DOS and carrier transport in a-Si:H.

electrons are thermally released later. Note that the density of the tail states near the valence band extends deeper into the bandgap. Hence, the holes trapped by these states need to wait longer for thermal release. Alternatively, they can hop to nearby states if their wavefunctions overlap. This mechanism of carrier transport is called hopping.

When electrons are trapped by the mid-gap states, thermal release is unlikely at room temperature. Negatively charged dangling bonds represent fixed charges that terminate electric field lines. Hence, electric field along the depth direction is no longer uniform. This is the reason why a bias is applied to a sample for a short period of time in the time-of-flight (TOF) experiment (see Section 2.1.1).

3.2 Low-temperature polycrystalline silicon

3.2.1 Film growth

Low-temperature polycrystalline silicon (LTPS) films are prepared on glass substrates by laser crystallization of a-Si films. A technique using an excimer

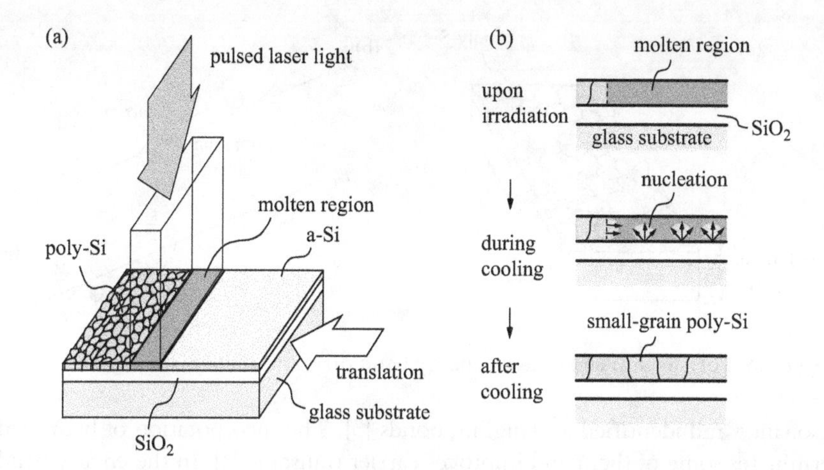

Figure 3.5 Crystallization of an a-Si film by laser annealing. (a) A very short pulse of laser light melts a linear region of the a-Si film. When it is cooled, a poly-Si film is formed. This process is repeated to crystallize the whole a-Si film. (b) As the molten Si is cooled, nucleation starts from multiple colder spots, resulting in a poly-Si film with small grains.

laser is called excimer laser annealing (ELA). Its conditions such as wavelength, power density, pulse width, and repetition rate, are carefully controlled to prevent thermal damage to the substrate. Otherwise, a deformed substrate would pose a problem for the following photolithography steps for fabricating TFTs.

Prior to laser crystallization, an a-Si film is deposited on a glass substrate. A thin SiO_2 layer is often inserted as an additional thermal barrier. Then, a pulsed laser light irradiates the linear region of an a-Si film as illustrated in Figure 3.5a. Incident optical energy is set high enough to melt the a-Si layer. The melting point of Si is 1,410°C. By limiting the downward heat flow with the SiO_2 layer, the substrate temperature is kept well below its glass transition temperature of around 600°C. Upon irradiation, at least a part of the a-Si film is melted. Heat flows from the molten region to colder regions. For example, when there are multiple colder spots on the substrate, heat flows vertically and multiple grains grow simultaneously as illustrated in Figure 3.5b. This condition results in a poly-Si film with small grains. By translating the substrate, one can repeat this crystallization process to form a large-area poly-Si film.

3.2.2 Electronic properties

As shown conceptually in Figure 3.6a, a polycrystalline material is composed of many small grains. Each grain is a single crystal and is randomly oriented. There are defects at the boundary of each grain. They can be detected by electron spin

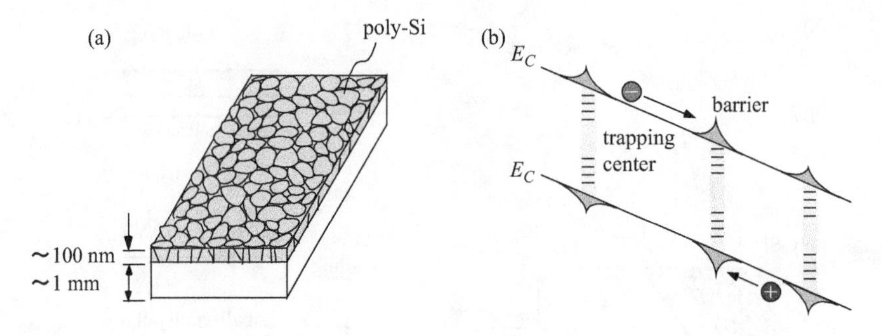

Figure 3.6 Poly-Si thin film and its energy band diagram under bias.

resonance and identified as dangling bonds [3]. The incorporation of hydrogen terminates some of them and improves carrier transport [4]. In the energy band diagram, grain boundaries are represented by potential barriers and states in the energy gap as shown in Figure 3.6b. Carriers need to go over the potential barriers (thermionic emission) or tunnel through (field emission). When trapped, they must be thermally released to become mobile again. In either case, carrier transport is hindered by these defects.

3.2.3 Lateral growth

By controlling heat flow, grains can be enlarged. For example, the energy density of the incident light is an important parameter that governs how nucleation proceeds. This dependency is illustrated in Figure 3.7. At low energy density, only the surface region is melted. Heat flows mainly toward the unmolten a-Si region as indicated by the arrows. Solidification starts

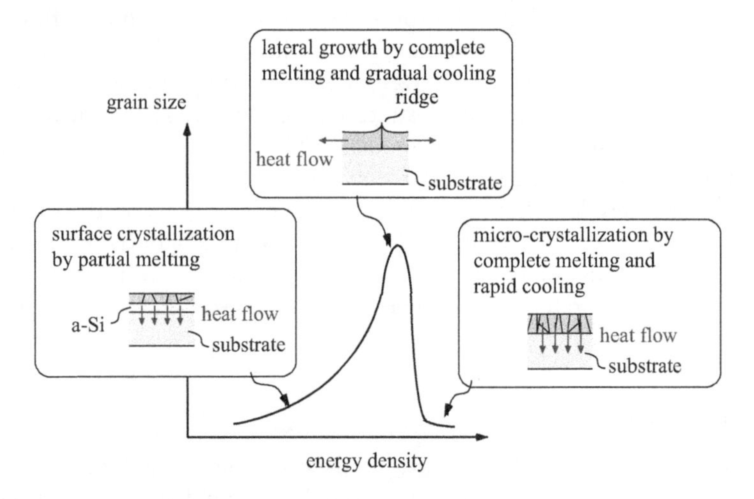

Figure 3.7 Control of grain size by the energy density of the incident light.

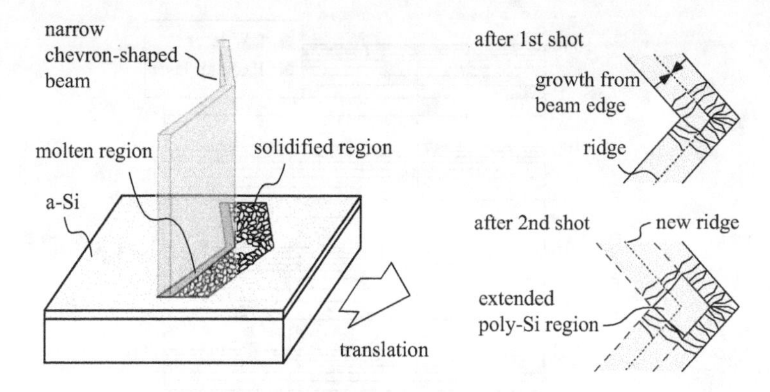

Figure 3.8 Crystallization with a chevron-shaped beam.

from numerous cold spots at the unmolten surface. This vertical solidification results in a small-grain film as described before. At higher energy density, the a-Si layer is completely melted. Heat flows through the molten silicon laterally, and solidification starts from the edges of the molten region. It completes when the two solidified regions collide with each other. As a result of lateral mass transfer, a ridge is formed in the film. Thus, lateral growth results in a film with large grains. When the incident energy density is increased further, vertical temperature gradient in the molten region becomes large. Heat flows mainly toward the substrate and this vertical solidification results in a small-grain film.

The ridges created during the laser annealing process can be problematic for TFT fabrication. Because TFTs are fabricated at a small region in each pixel of a display and an image sensor, one should be able to exclude these ridges. Alternatively, one can smoothen the film by repeatedly irradiating the film with an overlapping region. Effect of process conditions on grain size was systematically studied, and a grain size of 3 μm was reported [5].

Using chevron-shaped slits in place of a straight slit, a grain size of about 10 μm was reported for sequential solidification [6]. Its concept is illustrated in Figure 3.8. The first shot melts and solidifies a narrow chevron-shaped region. This results in a polycrystalline region with a ridge at the center. The second irradiation overlaps the first region. A relatively large polycrystalline region appears at the center. A new ridge is formed away from the first location. Subsequent shots extend the grain at the central region, while the ridge region keeps on shifting away from the enlarged grain region. Thus, a large-grain poly-Si film is obtained at the center of the chevron-shaped region.

3.2.4 Crystallization with continuous-wave (CW) laser

Crystallization with continuous-wave (CW) lasers was investigated before ELA. It is interesting to note that evolution of grain size realized by ELA

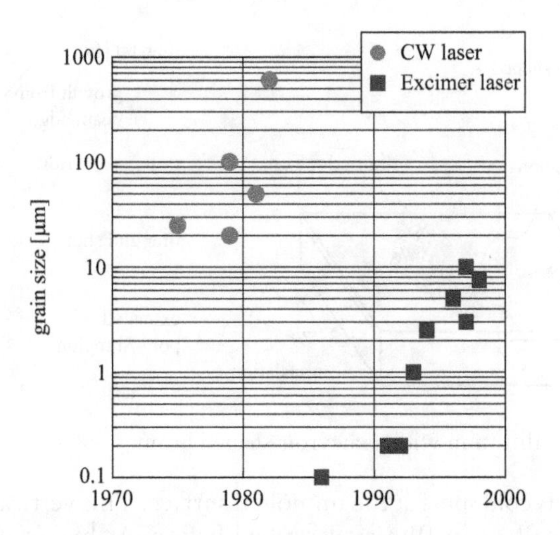

Figure 3.9 Evolution of grain size for CW laser and excimer laser crystallization techniques.

in the 1990s appears to follow the achievements by CW laser techniques as shown in Figure 3.9.

In 1975, with solar cell applications in mind, sputtered a-Si films with thicknesses ranging from 6 μm to 11 μm were crystallized by scanning a beam from a Nd:YAG laser [7]. The grain size was as large as 25 μm. The wavelength of this laser was 1.06 μm, and thick Si layers could be heated. Motivations of these works were not only solar cells but also TFT arrays and three-dimensional large-scale circuit integration. In 1979, it was reported that an Ar ion laser recrystallized 0.5 μm-thick poly-Si films deposited on silicon nitride layers. A narrow single crystal of 2 μm × 25 μm was obtained [8]. Poly-Si grain size reached 100 μm by crystallizing 0.5 μm-thick a-Si films deposited on a surface grating [9]. The authors named this technique graphoepitaxy. By shaping the laser beam to a half-moon shape, rectangular grains as large as 45 μm × 50 μm on quartz substrates were reported in 1981 [10]. Also in 1981, Biegelsen et al. patterned Si films to specify the point where nucleation starts. By scanning the CW Ar laser beam with elliptical intensity distribution, 20 μm-wide single-crystal regions were grown and extended [11]. In 1982, the use of a doughnut-shaped Ar laser beam increased the grain size to 600 μm [12].

These studies clearly show that heat flow and mass flow need to be controlled to grow a single crystalline film on an insulating layer by laser crystallization. Some important features were pointed out by Biegelsen et al. in 1981. First, by patterning Si films, one can specify the point where crystallization starts. For example, for the a-Si film patterned as illustrated in Figure 3.10, the temperature of the a-Si film drops first at the tip of

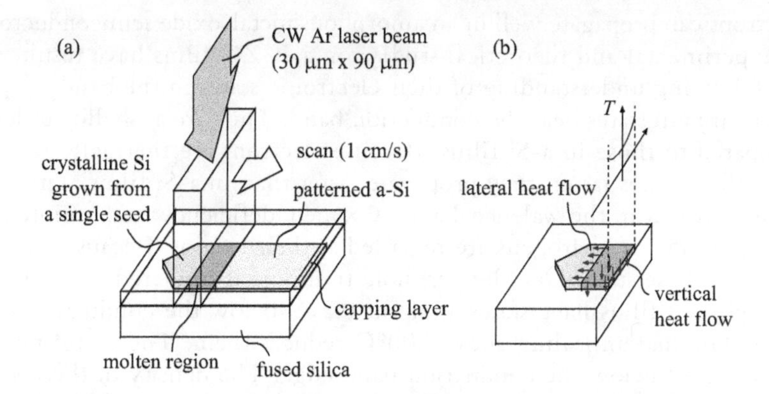

Figure 3.10 Extension of a single crystalline region by scanning a CW laser beam on a patterned a-Si film. (a) A crystalline Si grows from a single seed. (b) The temperature at the edges of the patterned film becomes higher when irradiated by a laser beam with an elliptical intensity profile.

the pattern. This is where nucleation starts. Second, nucleation from the edges of stripe-shaped molten region can be suppressed by controlling the intensity profile of a laser beam. For example, an elliptical intensity profile results in a slightly higher temperature at the edges of the patterned Si film. Third, a capping layer suppresses mass transfer during solidification. How to control the orientation of a single crystal remains a problem to be solved.

3.3 Metal-oxide semiconductor

3.3.1 Film growth

Metal-oxide films are usually deposited by sputtering at low temperatures, resulting in amorphous films. As-deposited amorphous InGaZnO (a-IGZO) films are thermally annealed at 300°C–400°C in oxidizing atmosphere. This post-deposition thermal treatment seems to be required for structural relaxation in these films. Adsorption and desorption of O_2, H_2, and H_2O are known to affect the electrical characteristics. In practical applications, protection layers are formed on TFTs to prevent these gas molecules from diffusing into active layers. In addition, irradiation of light can affect the TFT characteristics.

3.3.2 Electronic properties

High electron mobility in a-IGZO is attributed to the conduction path provided by spherical orbitals from metal atoms. Namely, in amorphous oxide containing metal atoms, spherical orbitals of the neighboring metal atoms overlap, providing a conduction path. The magnitude of this overlap is insensitive to the orientation of the metal-oxygen-metal bond. Hence, electrons can propagate well in an amorphous metal-oxide semiconductor.

Experimental and theoretical studies on a-IGZO films have resulted in the following understanding of their electronic states in the bandgap [13]. There are tail states near the conduction band. They are at shallower levels compared to those in a-Si films. Because electrons are thermally released from these states faster, they propagate faster than in a-Si films. There are deep states near the valence band. Oxygen deficiency and unintended incorporation of hydrogens are regarded as their origins. Because hopping between these states takes longer, hole transport is hindered. In addition, as-deposited films have states at about 0.2 eV below the conduction band edge. Thermal annealing above 300°C reduces them. There are states at about 0.3 eV below the conduction band edge. The density of these states decreases by incorporation of hydrogen, and it increases by annealing above 400°C.

References

1. P. G. Le Comber and W. E. Spear, "Electronic transport in amorphous silicon films," *Phys. Rev. Lett.* 25, 509–511 (1970).
2. R. A. Street, J. Zesch, and M. J. Thompson, "Effects of doping on transport and deep trapping in hydrogenated amorphous silicon," *Appl. Phys. Lett.* 43, 672–674 (1983).
3. N. M. Johnson, D. K. Biegelsen, and M. D. Moyer, "Deuterium passivation of grain-boundary dangling bonds in silicon thin films," *Appl. Phys. Lett.* 40, 882–884 (1982).
4. N. H. Nickel, N. M. Johnson, and W. B. Jackson, "Hydrogen passivation of grain boundary defects in polycrystalline silicon thin films," *Appl. Phys. Lett.* 62, 3285–3287 (1993).
5. A. Marmorstein, A. T. Voutsas, and R. Solanki, "A systematic study and optimization of parameters affecting grain size and surface roughness in excimer laser annealed polysilicon thin films," *J. Appl. Phys.* 82, 4303–4309 (1997).
6. J. S. Im, R. S. Sposili, and M. A. Crowder, "Single-crystal Si films for thin-film transistor devices," *Appl. Phys. Lett.* 70(25), 3434–3436 (1997).
7. J. C. C. Fan and H. J. Zeiger, "Crystallization of amorphous silicon films by Nd:YAG laser heating," *Appl. Phys. Lett.* 27, 224–226 (1975).
8. J. F. Gibbons, K. F. Lee, T. J. Magee, J. Peng, and R. Ormond, "cw laser recrystallization of <100> Si on amorphous substrates," *Appl. Phys. Lett.* 34, 831–833 (1979).
9. M. W. Geis, D. C. Flanders, and H. I. Smith, "Crystallographic orientation of silicon on an amorphous substrate using an artificial surface-relief grating and laser crystallization," *Appl. Phys. Lett.* 35, 71–74 (1979).
10. T. J. Stultz and J. F. Gibbons, "The use of beam shaping to achieve large-grain cw laser-recrystallized polysilicon on amorphous substrates," *Appl. Phys. Lett.* 39(6), 498–500 (1981).

11. D. K. Biegelsen, N. M. Johnson, D. J. Bartelink, and M. D. Moyer, "Laser-induced crystallization of silicon islands on amorphous substrates: Multilayer structures," *Appl. Phys. Lett.* 38, 150–152 (1981).
12. S. Kawamura, J. Sakurai, M. Nakano, and M. Takagi, "Recrystallization of Si on amorphous substrates by doughnut-shaped cw Ar laser beam," *Appl. Phys. Lett.* 40, 394–395 (1982).
13. K. Ide, K. Nomura, H. Hosono, and T. Kamiya, "Electronic defects in amorphous oxide semiconductors: A review," *Phys. Status Solidi A* 216, 1800372 (2019).

4 Image sensors

In an image sensor for visible light, an array of photodetectors converts an incident radiation intensity pattern to a distribution of electric charges. Depending on how to read out these charges, there are two technologies: metal–oxide–semiconductor (MOS) image sensor [1] and charge-coupled device (CCD) [2]. Photogenerated charges are read out through metallic lines in a MOS image sensor whereas they are transferred step by step in the vicinity of the oxide–semiconductor interface in an early CCD. Although the MOS image sensor was conceived earlier than CCD, the latter dominated the market for many years due to its superior image quality. Active-pixel sensor technology [3] changed this in the 2010s. It is now called complementary MOS (CMOS) image sensor and is widely put in consumer products such as digital cameras and smartphones. Combined with various optical components, they are applied for scanners, facsimiles, and copiers as well. In addition to visible light, capacitance, heat, and pressure can be detected for acquiring fingerprint images.

Readout scheme of MOS image sensors is also utilized for capturing X-ray and infrared images. Because photodiodes based on Si are almost transparent to these radiations, other schemes are required. For X-ray imaging, a scintillator is attached to a photodiode array to convert X-rays to visible photons. Alternatively, amorphous selenium converts X-rays to electron–hole pairs directly. In both cases, a large-area readout circuit is required to cover a human body for medical imaging or a suitcase for non-destructive inspection. Thin-film transistors provide a solution to this problem [4]. For infrared imaging, there are two configurations as well. Compound semiconductors with small energy bandgaps and crystalline Si with impurity bands can convert an infrared photon to an electron–hole pair directly. These materials need to be cooled to suppress thermal noise. Alternatively, one can detect temperature rise caused by absorption of infrared photons with devices based on micro-electro-mechanical system (MEMS) technology. These imagers do not require cooling [5].

In this chapter, image sensors for visible light, X-rays, and infrared light are described in this order. Common approaches are often adopted for improving characteristics of image sensors. For example, pixel-level amplification enhances signal-to-noise ratio (SNR) and monolithic integration reduces manufacturing costs.

DOI: 10.1201/9781003364979-4

4.1 MOS image sensors

In the history of solid-state image sensors, the late 1960s was an exciting period because three fundamental concepts were reported. They are amplification of photogenerated charges at each pixel, charge storage mode operation, and buried photodiodes. These rationales set goals for later developments. Current CMOS image sensors adopt all of them. These concepts are described by referring to the examples illustrated in Figure 4.1. In all cases, signal charges are transferred from each pixel to external circuits through global paths made of highly conductive materials such as metals and heavily doped semiconductors.

In 1966, Schuster and Strull reported the scheme shown in Figure 4.1a [6]. A bipolar transistor is used to address a pixel. It also functions as a photo-detector with an internal gain mechanism. Although it is not a MOSFET, this configuration is included here because of the common feature for addressing each pixel. Two global lines are connected to the collector and emitter of the transistor at each pixel. Its base is floating. This configuration was implemented on crystalline Si substrates. Incoming light generates electrons and holes in the p-n junctions of the bipolar transistors. The horizontal and vertical shift registers connect only one of them to an external circuit at a time. The number of electrons collected by the external circuit is larger than the number of incident photons in the short period of addressing time. Therefore, the signal is amplified at each pixel. By repeating this process for each pixel sequentially, a video signal is generated. However, photogenerated carriers in unaddressed transistors are lost via recombination at the p-n junctions. Hence, the photons incident on most pixels do not contribute to signal generation.

In 1967, Weimer et al. reported the scheme shown in Figure 4.1b [7]. A photoconductor converts incident light to charges, which reduces its resistance in turn. The horizontal and vertical shift registers turn on the MOSFET in

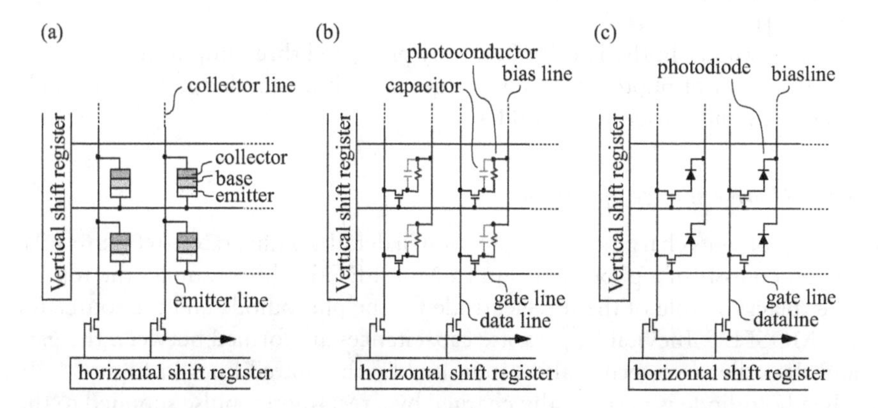

Figure 4.1 Three architectures for MOS image sensors. The photodetector in a pixel is (a) phototransistor, (b) photoconductor, and (c) photodiode.

each pixel one after another to read out signals. The signal in terms of the number of electrons is larger than the number of incident photons because electrons are injected from the contact of the photoconductor. Again, this is a signal amplification process at each pixel. There is another important feature in this scheme. The capacitor across the photoconductor enables the so-called integrating mode operation. It is also called charge storage mode. When a pixel is addressed, its capacitor is fully charged. Incident light reduces the resistivity of the photoconductor. While the MOSFET is turned off, photogenerated charges cancel the charges stored in this capacitor. The time constant for this discharge is proportional to the resistivity of the photoconductor. When this pixel is addressed next time, the charges supplied to the capacitor are proportional to the photogenerated charges. The time interval between successive addressing is called integration time. Because almost all photogenerated charges contribute to signal generation, this operation mode greatly improves sensitivity of an image sensor. Note that the photoconductor material was CdS and that the pixel switch was a CdSe TFT fabricated on glass substrates. A year later, Sadasiv et al. integrated CMOS shift registers with CdSe TFTs [8].

In 1967, Weckler reported a one-dimensional version of the configuration shown in Figure 4.1c. Each pixel consisted of a crystalline Si photodiode and a MOSFET [9]. Although a photodiode does not provide an internal amplification mechanism, its output is proportional to the number of incident photons for a larger range of photon flux than a sensor based on photoconductivity. This is an important feature for acquiring high-quality images. Because a photodiode is a capacitor, the pixel configuration for realizing the integrating mode is simplified. In 1968, Noble reported a two-dimensional version [10]. In retrospect, two valuable features were also noted in this paper. First, a photodiode is connected to the gate terminal of a MOSFET. This is a pixel circuit for signal amplification. Second, the merit of buried photodiodes was mentioned. They decrease leakage current. Buried photodiodes are also called pinned photodiodes. In 1982, they were implemented in CCDs [11] and later adopted by CMOS image sensors [12].

These studies in the late 1960s clearly presented three important concepts: amplification of photogenerated charges at each pixel, charge storage-mode operation, and buried photodiodes.

4.1.1 Charge storage operation

Let us look at its charge storage operation in detail. As illustrated in Figure 4.2a, a pixel consists of a photodiode and a MOSFET. In this example, the top n+ layer plays the role of the top electrode for the photodiode and the source for the MOSFET. Inevitably, parasitic capacitances are formed between the gate and the other two terminals as indicated by the dotted lines in Figure 4.2b. The photodiode is periodically charged by a rectangular pulse supplied to the gate terminal of the MOSFET. The time constant for this charging process is

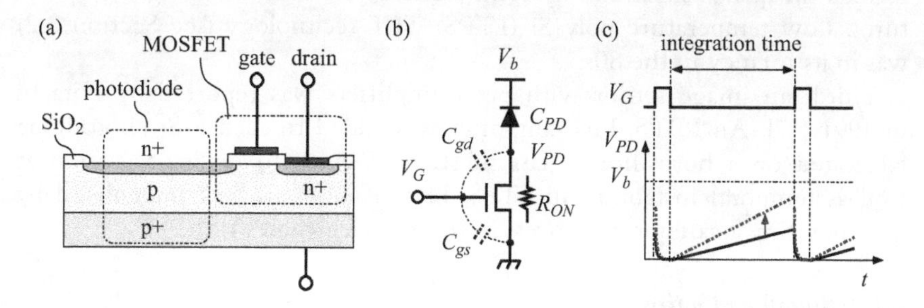

Figure 4.2 Operation of MOS image sensor. (a) Cross section of a pixel in a crystalline Si MOS image sensor, (b) its equivalent circuit, and (c) timing chart.

approximately $R_{ON}C_{PD}$, where R_{ON} is the on-time resistance of the MOSFET and C_{PD} is the photodiode capacitance. Assuming that $R_{ON}C_{PD}$ is sufficiently short compared to the gate on-time, the photodiode is fully charged to $Q_o = C_{PD}V_b$, where V_b is the bias voltage. Under constant illumination, photogenerated charges cancel Q_o at a fixed rate. Hence, the potential at the node between the photodiode and the MOSFET (V_{PD}) increases with time as shown in Figure 4.2c. At a larger incident photon flux, V_{PD} increases more rapidly as shown by the dotted line. After a certain fixed duration, the pixel is addressed again. The time duration when the MOSFET is at off-state is called integration time. The charges flowing into the photodiode are recorded by an external circuit connected to this node. This is the signal proportional to the photon flux incident on the photodiode. Note that the maximum signal that can be detected is equal to $Q_o = C_{PD}V_b$.

4.1.2 Large-area MOS image sensors

In the 1980s, hydrogenated amorphous silicon (a-Si:H) was extensively studied for developing solar cells to harvest energy from sunlight and TFTs to drive liquid crystal displays. Other potential applications included photosensitive drums for printers and copiers and contact-type MOS image sensors for facsimiles and scanners. Photosensors and TFTs were fabricated with a-Si:H on glass substrates. Although CdS-CdSe was employed in 1981 [13], various a-Si devices were investigated later. There are some notable features in these activities as follows.

First, one can form a-SiC layers by mixing methane gas in the PECVD process. This wide bandgap material improves the quantum efficiency of a photodiode at short wavelengths when it is used at its entrance side. It also reduces the dark current. Heterojunction configurations such as p-type a-SiC/intrinsic a-Si/a-SiN_x [14] and p-type a-SiC/intrinsic a-Si/ n-type a-SiC [15] were reported.

Second, driver circuits were integrated on the same substrates with poly-Si TFTs [16]. This architecture dramatically reduced the number of connections to external circuits. In this study, however, poly-Si layers were formed on quartz substrates by solid-phase crystallization at high temperatures. Low-temperature poly-Si (LTPS) TFT technology (see Section 3.2) was in its infancy in the 80s.

Third, an image sensor with pixel amplifiers was reported by Hitachi in 1991 [17]. An LPTS TFT amplifier was added to each a-Si photodiode fabricated on a borosilicate glass substrate. Device to device variation as well as temporal instability in LTPS TFTs if any can be compensated by incorporating a calibration process in readout electronics [18].

4.1.3 Signal and noise

When a photon flux is incident on an image sensor at a constant rate, its output signal is proportional to the photon flux and the integration time. The maximum signal charge is equal to the product of the photodiode capacitance and its bias ($Q_o = C_{PD}V_b$). SNR is defined as the ratio of this number and the noise. Therefore, it is critically important to reduce the noise to acquire a high-quality image. There are two types of noise: fixed pattern noise (FPN) and random noise.

At the leading edge and the trailing edge of the gate pulse, charges are injected into the photodiode through the parasitic capacitance C_{gd}. Variations in the fabrication process cause C_{gd} to vary from pixel to pixel. Hence, a fixed pattern is added to the signal waveform. This is called fixed pattern noise (FPN). One can eliminate it by image processing techniques or cancel it with dedicated circuits that record these charges [19]. However, it still sets a lower limit on the photon flux that can be detected. Denoting the gate voltage swing as ΔV_g, the charge injected through C_{gd} is given by $C_{gd}\Delta V_g$. For example, one can minimize it by reducing C_{gd} with a self-aligning technique and/or reducing ΔV_g with low-threshold processes.

Random noise is more serious because it fluctuates with time. First, electrons in a conductor are scattered by lattice vibrations and this results in a voltage fluctuation across the conductor. This is called thermal noise. Because the channel of a MOSFET is a conductor, it generates thermal noise. Noise voltage variance of a resistor R at temperature T is modeled by a voltage source $v_n^2 = 4kTR\Delta f$, where f is the frequency, Δf is the bandwidth, and k is Boltzmann's constant. When a capacitor is connected to the resistor in series, a part of the noise voltage drops across the capacitor. Hence, the following equation holds for the noise voltage across the capacitor V_n.

$$V_n^2 = \int_0^\infty \left| \frac{\dfrac{1}{i2\pi fC}}{R + \dfrac{1}{i2\pi fC}} \right|^2 \times 4kTR\,df = 4kTR \int_0^\infty \frac{1}{1+(2\pi RCf)^2} df = \frac{kT}{C} \quad (4.1)$$

The fluctuation in the charge stored on the capacitor C connected to a resistor at temperature T is equal to $CV_n = \sqrt{kTC}$. For this reason, it is also called kTC noise.

In addition, there is a noise called 1/*f* noise or flicker noise. Its physical origins are not clear. Possibilities argued so far include fluctuations in the mobility of the carriers in the MOSFET channel, and carrier generation and recombination from the traps at the insulator-semiconductor interface in a MOSFET [20]. Empirically, its power spectrum is inversely proportional to the frequency. This is where the term comes from. In contrast, thermal noise is called white noise because it does not depend on the frequency.

Finally, the number of photons incident on a pixel follows the Poisson distribution. So does the number of electrons flowing in a circuit. If this number is N, its statistical fluctuation is $\sqrt{N}$. This is called shot noise.

4.1.4 Pixel-level amplification

As the studies in the past clearly indicate, a photodiode is a preferred detector over a photoconductor because of its wide range of linear input-output characteristics. However, its quantum efficiency never exceeds unity. Amplification of signal charges in a photodetector itself is challenging. Alternatively, it can be accomplished by adding an amplifier to each pixel. This concept was pointed out by Noble et al. for MOS image sensors on Si substrates in 1968. For infrared imaging, it was demonstrated with a two-dimensional array of indium antimonide (InSb) detectors in 1987 [21]. Each detector was connected to a source follower circuit on a Si substrate through an indium bump. This is an example of a hybrid structure (see Section 4.5.1). For radiation detector applications, some preliminary designs of pixel amplifiers were reported in the late 80s (see Section 4.4.1). For contact-type linear image sensors, Hitachi demonstrated pixel-level signal amplification with a LTPS TFT circuit connected to an a-Si photodiode in 1991 (see Section 4.1.2).

In 1993, Eric R. Fossum pointed out several problems of CCD technology for space missions: radiation softness, small area sizes, complexity for CMOS circuit integration, etc. The term "active-pixel sensor" was invented for a MOS image sensor with pixel-level signal amplification [22].

An example configuration is shown in Figure 4.3. Each pixel consists of four MOSFETs, a photodiode, and a capacitor. The column circuit can include an amplifier, sample-and-hold circuits, and even an analog-to-digital converter. A signal is read out from the pixel as follows. First, the gate terminal of the transistor SF is reset to a fixed bias by turning on the transistor RST. Second, the transistor TG is turned on to connect the gate terminal of the transistor SF to the photodiode. Third, the transistor RS is turned on to let the signal current flow into the column circuit. The signal charges remain in the capacitor while the transistor RST is turned off. Therefore, a non-destructive readout is possible. This pixel circuit configuration, together with sample-and-hold circuits in the column circuit, allows cancellation of *kTC* noise from the transistor RST by the technique called correlated double sampling (CDS) as follows. There are two sample-and-hold circuits in the column circuit. They hold the voltage on the column metal line before and

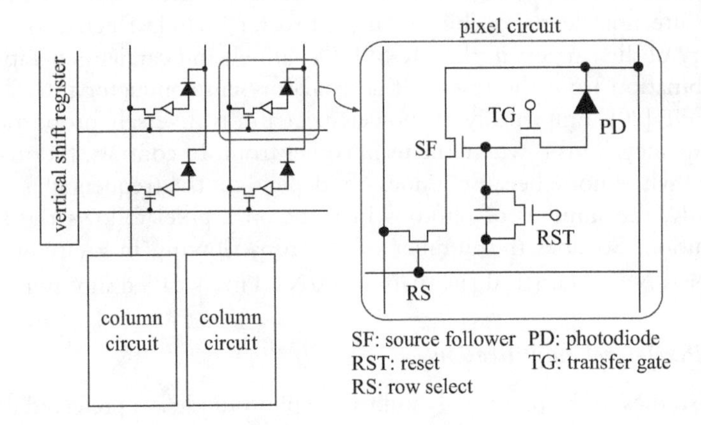

Figure 4.3 An example of the readout circuit configuration of an APS.

after the photodiode is connected to the capacitor in the pixel. The voltage sampled before the signal charge transfer contains the noise voltage from the transistor RST. Therefore, subtracting one from the other cancels the *kTC* noise from the transistor RST. This noise cancellation technique has been adopted by CCDs [23] and CMOS image sensors [24].

In the case of linear image sensors, plenty of space is available near a photodiode to lay out the pixel circuit. Obviously, this is not the case for two-dimensional image sensors. In the conventional design called front-side illumination, a photodiode and pixel circuit share a limited pixel area. The ratio of a photosensitive area and a pixel area is called fill factor. A large fill factor is desired to increase signal charges. An on-chip microlens array focuses incoming light on photodiodes to increase light utilization. For reducing the cost of consumer electronics, minimizing die size is effective. The cost of optics is also reduced. However, pixel counts need to be large for acquiring high-resolution images. Hence, pixel pitch must decrease. Thanks to the steady reduction of minimum feature size in the integrated circuit industry, CMOS image sensor technology had become a serious competitor to CCDs in the 1990s [25].

Sharing some MOSFETs with neighboring pixels is effective for decreasing the area occupied by the circuit. In principle, MOSFETs except one addressing each photodiode (labeled as TG in the figure) can be shared. For example, sharing MOSFETs with four pixels resulted in 1.5 transistors per pixel in 2004 [26]. In 2007, pixel pitch and transistor count per pixel reached 1.75 μm and 1.75/pixel, respectively [27]. In 2011, these numbers reached 1.12 μm and 1.25/pixel, respectively [28]. A photograph of eight pixels sharing three MOSFETs is found in Ref. [29]. Typical fill factor for these front-illuminated devices is below 50% [30]. It can be close to 100% for a back-illuminated device [31] although using a silicon-on-insulator (SOI) wafer or thinning a Si wafer adds cost. Incidentally, the concept of backside illumination was applied to CCDs in 1974 [32].

Pixel shrinkage cannot continue forever because of physical limits. Shot noise would deteriorate image quality: the number of photogenerated electrons fluctuates due to statistical variation in the number of photons incident on a photodiode. Because absorption coefficient is a material property, some obliquely incident photons would be absorbed by neighboring photodiodes, resulting in cross-talk. Diffraction starts to limit the benefit of microlens as the pixel size shrinks beyond 2 μm [33].

4.1.5 Monolithic integration

One can fabricate a photodetector on top of a substrate with pixel circuits as illustrated in Figure 4.4. An insulating layer is inserted between them for planarization. Via holes connect the photodetector to each pixel circuit. In fact, X-ray and some infrared imagers adopt this configuration (see Sections 4.4 and 4.5). In theory, a fill factor can reach 100% with this approach.

As for the choice of the photodiode and MOSFET technologies, various combinations are possible. For example, an image sensor based on a-Si photodiodes and LTPS TFTs was reported in 2002 [34]. For pixel-level signal amplification, three LTPS TFTs (a reset switch, a source follower, and a pixel selection switch) were fabricated in a 90 μm × 90 μm pixel area. A p-i-n a-Si photodiode was fabricated on top of these TFTs by PECVD and photolithography. Also, in 2002, a continuous organic photoconversion layer was formed on an a-Si TFT array [35]. The sensor consisted of a photogeneration layer and a charge transport layer. The two layers were prepared by either evaporation or solution processes. The materials were all small molecules and carriers were transported through their π-orbitals via hopping. In 2008, a uniform organic sensor was integrated into an a-Si TFT array by solution processes [36]. This a-Si TFT array was fabricated on a polymer substrate at less than 150°C. In these imaging demonstrations, a-Si TFT arrays without pixel amplifiers were used.

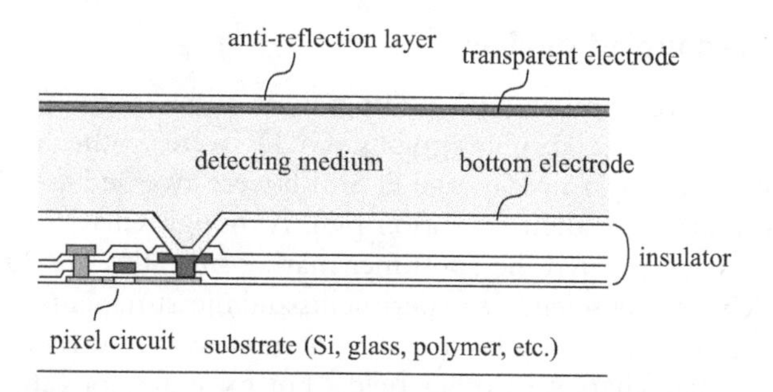

Figure 4.4 Cross section of a detector monolithically integrated on a substrate with readout circuits.

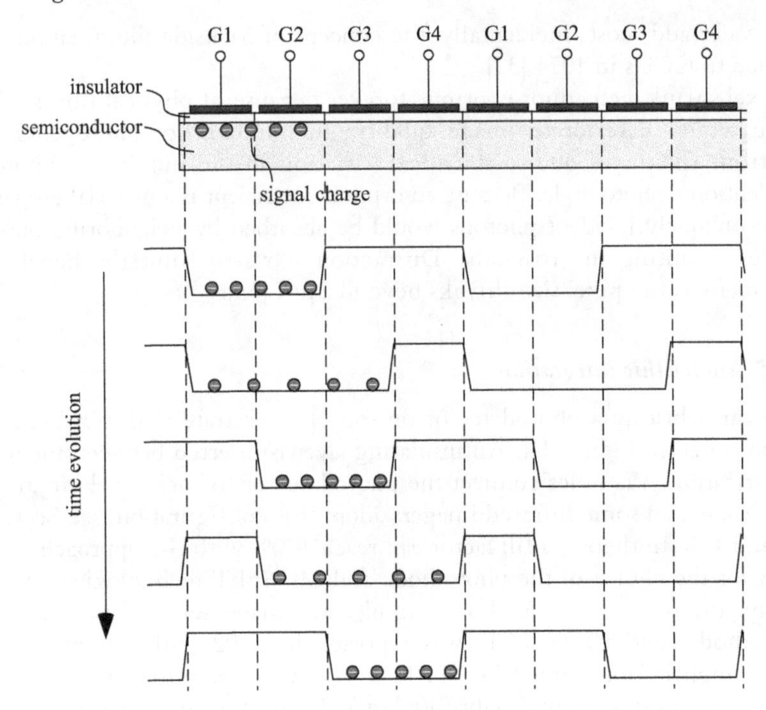

Figure 4.5 Charge transport in a CCD.

Integration of colloidal quantum dots on CMOS readout circuits is described in a review paper published in 2022 [37]. Although their objective is capturing short-wavelength (1.4 μm–2.5 μm) infrared images, some groups report quantum efficiencies exceeding 60% in the visible range. Bandgap energy of quantum dots can be tuned. Because signal size of an image sensor is proportional to the product of fill factor and quantum efficiency, a high fill factor realized by monolithic integration might be able to overcome an inferior quantum efficiency.

4.2 Charge-coupled devices

For those who are familiar with photographic films, digital cameras have changed the way we take photographs. CCDs were at the heart of these products. Willard Boyle and George E. Smith were awarded the Nobel Prize for Physics in 2009 for their invention [38]. Although CMOS image sensor technology has dominated the consumer market since the late 2010s, CCDs are still widely used in scientific experiments and measurements. The mechanism of charge transfer in CCDs might be of interest to researchers who wish to control electric charges in other fields. For example, ions in electrolytes might be manipulated by multiple electrodes in microfluidic chip systems for biosensing applications.

4.2.1 Basic configuration and operation

A CCD consists of closely packed MOS capacitors and metal wirings. Charges are transferred by applying appropriate biases to their metal electrodes. An example of voltage waveforms is shown in Figure 4.5. When a positive bias is applied to the electrodes G1 and G2, a potential well is created in the semiconductor region under them. Photogenerated carriers (electrons in this case) are accumulated at the SiO_2-Si interface under G1 and G2. Next, a positive bias is applied to G3, and the potential well extends under G3. The electrons spread under the three electrodes. When the bias on G1 is turned off, the potential well shrinks. Now, the electrons are under G2 and G3. By repeating this process, charges are transferred at the SiO_2-Si interface.

One might wonder where the charges come from. Light cannot pass through a metal layer unless it is extremely thin. If the top layer of the MOS capacitor is made of heavily doped poly-Si, some light goes through it. Electrons and holes are generated in the semiconductor via the photoelectric effect. Thus, a CCD itself functions as a photodetector. Even though the poly-Si layers are thin, they absorb photons, especially at short wavelengths. A better configuration is to add a separate photodiode as will be explained below.

Three types of CCDs are developed for two-dimensional imaging. A full-frame CCD is shown in Figure 4.6a. All MOS capacitors convert light to charges and store them. These charges are transferred by vertical CCDs row by row to the horizontal CCD, which outputs a serial signal to an external circuit. A mechanical shutter prevents charge generation, while one image called frame is read out. Nevertheless, a mechanical component is prone to

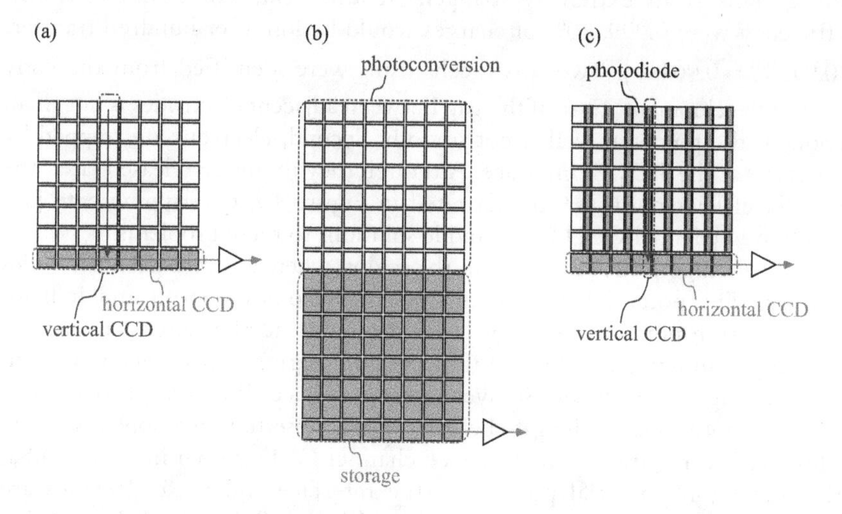

Figure 4.6 Three architectures for two-dimensional CCDs: (a) full-frame CCD, (b) frame-transfer CCD, and (c) interline-transfer CCD. The shaded regions represent MOS capacitors with a light shield.

wear. The second design shown in Figure 4.6b is called frame-transfer CCD. As shown by the shaded regions, it has an additional CCD with light shields. The upper CCD is used for photoconversion. The charges stored in the upper CCD are rapidly transferred to the lower CCD with light shields. While this frame is read out by the lower CCD, the upper CCD generates the next frame. In fact, the upper CCD converts the incoming light to charges even while a frame is being transferred. When intense light falls on one pixel, charges flow to neighboring pixels in the vertical CCD. This results in a white vertical line in an image. The artifact is called smear and is inevitable in a frame-transfer CCD. To prevent this effect and to increase the efficiency of photoconversion, typical digital cameras adopt an architecture called interline CCD shown in Figure 4.6c. A photodiode and a MOSFET are added to each pixel. Electrons generated by the photodiode are transferred to the vertical CCD through the MOSFET. A full frame is read out in the same manner as the scheme shown in Figure 4.6a. A microlens array is fabricated on it to focus incoming light on the photodiodes.

Incidentally, the term "CCD" is sometimes used for the device that transfers charges only. It is also used for the device with both functions of photoconversion and charge transfer. Furthermore, it sometimes means a CCD image sensor.

4.2.2 Enhancement of transfer efficiency

Because signal charges are transferred so many times in a CCD, transfer efficiency from one MOS capacitor to another must be very close to unity. This is an extremely stringent requirement. For example, if this efficiency were 0.999, 10% of charges would be lost after hundred transfers $\left(0.999^{100} = 0.904\cdots\right)$. Two loss mechanisms were identified from the early days of development. First, if the gap between adjacent electrodes is not small enough, the potential well is not smooth. Second, electrons are trapped by the states at the SiO_2-Si interface. Although they might be released later, the transfer efficiency drops. As illustrated in Figure 4.7, overlapping gate [39] and buried channel CCD [40] provide solutions to these problems.

First, a fringing field develops in the region where adjacent gate electrodes overlap. The potential well in this region becomes smooth and it helps electron transport. Second, electron trapping at the interface can be avoided by moving the channel into the bulk region. This is accomplished by inserting a thin doped Si layer at the interface. Potential distributions under a positive bias at the gate before and after insertion of a doped layer are illustrated in Figure 4.8. In a surface channel CCD shown in Figure 4.8a, the minimum potential point is at the interface and most electrons are transferred in this region. They are more likely to be trapped. In a buried channel CCD shown in Figure 4.8b, this point moves into the bulk region. Because trapping is much more unlikely, the transfer efficiency is

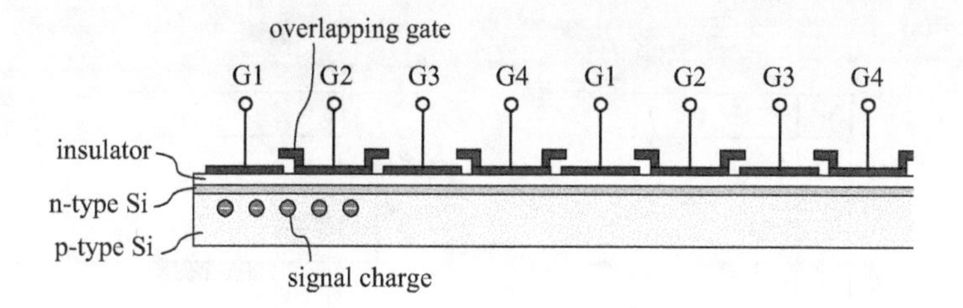

Figure 4.7 Cross section of a buried CCD with overlapping gate electrodes.

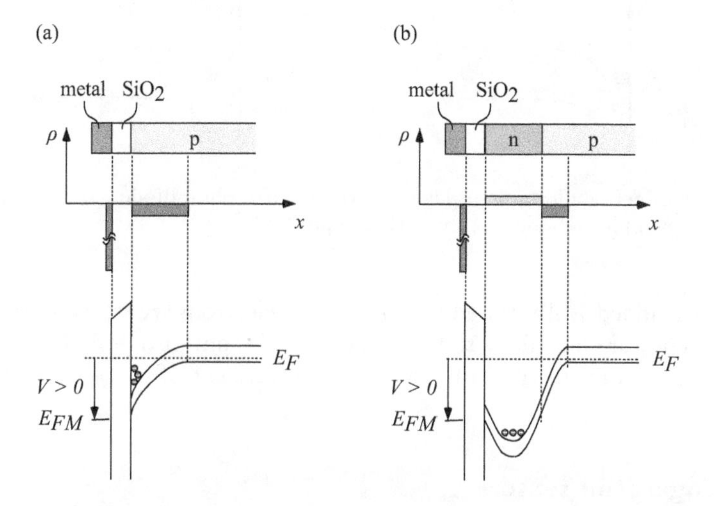

Figure 4.8 Effect of inserting an n-type layer between the oxide and the p-type Si. (a) In a surface channel CCD, electrons are accumulated at the interface with the insulator. (b) In a buried channel CCD, they are transferred in the bulk region.

enhanced. The presence of stored charges modifies these potential distributions. One-dimensional electrostatic model was used to calculate potential distributions of a buried channel structure [41]. Of course, the two solutions are independent; i.e., the concepts of overlapping gate and buried channel can be combined [42].

4.2.3 Buried photodiodes

In an interline-transfer CCD, the signal charge is transferred from a photodiode to a vertical CCD through a MOSFET. In a conventional p–n photodiode, most electrons are stored in the region near the surface of the n-type Si as shown in Figure 4.9a. This is where traps are present. Trapped electrons are released later and these add background to an image. When a

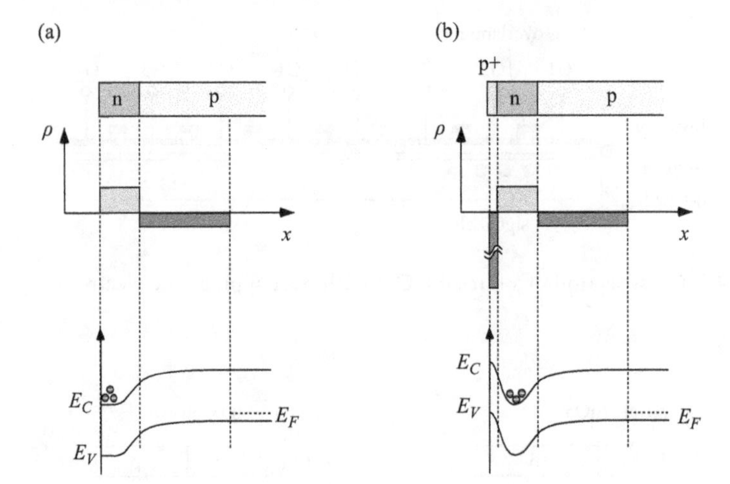

Figure 4.9 Effect of adding a p+ layer on top of a p-n photodiode. (a) Conventional p-n photodiode and (b) buried photodiode.

p+ layer is added as illustrated in Figure 4.9b, electrons are stored in the bulk region. They are less likely to be trapped in this configuration. The concept of storing charges in the bulk region is reminiscent of the buried channel CCDs.

4.3 Fingerprint sensors

Optical fingerprint sensors detect visible light reflected and/or scattered by the ridges and valleys of a finger. Other features arising from the ridges and valleys, such as capacitance, heat flow, and pressure, can be detected to acquire fingerprint images. Optical sensors are explained first. Description of sensors based on other mechanisms follows.

4.3.1 Optical sensors

Visual inspection of a finger immediately reveals that the contrast of a fingerprint image is low. One can enhance it optically. For example, a finger in contact with the hypotenuse of a prism is illuminated by collimated light as illustrated in Figure 4.10. The illumination direction and the refractive index of the prism are chosen such that the condition for total internal reflection (TIR) is satisfied at the hypotenuse. Note that air gaps exist at the valleys of the finger. The light incident on the valley regions is reflected and enters the camera. The ridges scatter the incident light, and the scattered light barely reaches the camera. Thus, the valleys are enhanced in the image recorded by the camera. In the original configuration reported in 1962, a viewing screen was used instead of a camera [43].

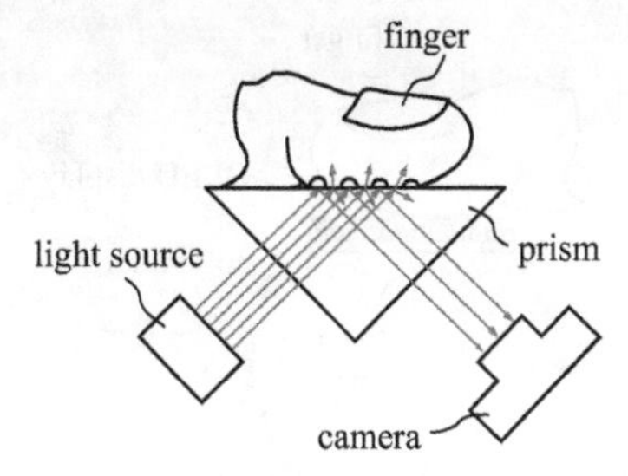

Figure 4.10 Contrast of a fingerprint image is enhanced by TIR at the valleys of a finger.

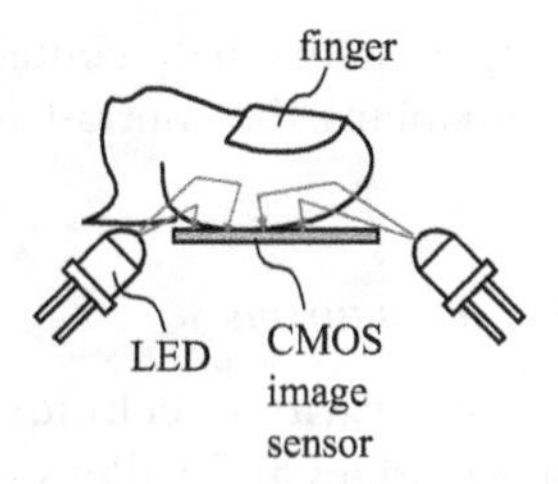

Figure 4.11 A CMOS image sensor in contact with a finger detects the light scattered inside.

Inevitably, the thickness of this configuration exceeds that of the prism. The invention of MOS image sensors and CCDs has enabled contact-type imaging. For example, a finger in contact with a CMOS image sensor is illuminated by the LEDs placed around it as illustrated in Figure 4.11 [44]. The light enters the finger and propagates inside. The scattered light emerges from the finger and is detected by the image sensor.

The central region of the finger is not well illuminated in this configuration. This problem is solved by a two-dimensional image sensor fabricated on a transparent substrate by TFT technology. A planar light source such as a backlight unit for an LCD is stacked beneath it. The light passing through the transparent regions of the image sensor illuminates the finger. The reflected light is detected by the image sensor in contact with the finger.

Furthermore, a fingerprint sensor integrated with a display was reported in 2019 [45]. As illustrated in Figure 4.12, a top-emission OLED display (see Section 5.2) is stacked on a collimator and an image sensor. The OLED display is fabricated on a transparent substrate. Light passes through the regions not occupied by OLEDs and metal lines. When the OLED display illuminates the finger, the reflected light passes through the transparent regions of the OLED display and enters the collimator. Because the light propagating toward neighboring pixels is absorbed by the collimator, degradation of spatial resolution is prevented.

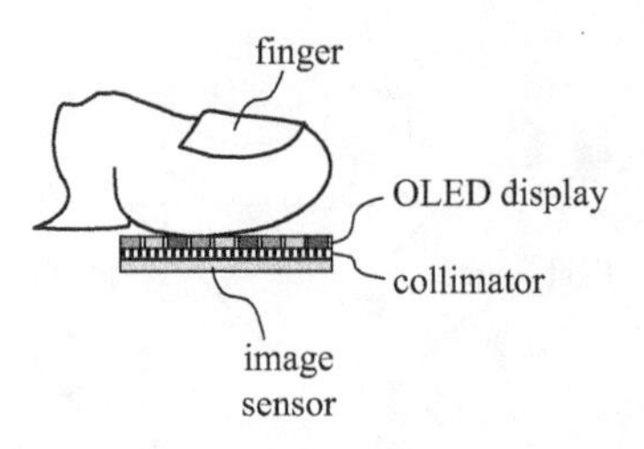

Figure 4.12 An OLED display can be used to illuminate a finger for acquiring its image.

Unlike the sensor utilizing a prism, these contact-type optical sensors do not have a mechanism for enhancing the contrast of the ridges and valleys.

4.3.2 Sensors based on other mechanisms

In addition to optical sensors, various techniques have been developed for converting the pattern of ridges and valleys on the surface of a finger to electric charges on each pixel of an image sensor. Three examples are illustrated in Figure 4.13.

In Figure 4.13a, an array of thermal sensors is in contact with a finger through a thin protective layer. They detect temperature distributions of a finger. The air gaps at the valleys of a finger represent heat resistance. Thus, heat exchange at the valleys is smaller than at the ridges. Because thermal equilibrium is established quickly, an image acquired by a two-dimensional sensor would disappear. To cope with this problem, a finger is swiped on a linear array of thermal sensors.

In Figure 4.13b, capacitive coupling between the finger and pixel electrodes is utilized for imaging. The distance between the finger and the pixel electrodes is larger at the valleys. Hence, the capacitance is smaller for the pixel electrodes below the valleys. This distribution is detected by applying pulsed bias on the pixel electrodes. A capacitive fingerprint sensor based on LTPS technology was reported in 2003 [46]. Because the presence of moisture in a finger affects capacitances, dry and wet fingers result in different images.

In Figure 4.13c, MEMS technology is applied to form cavities between upper and lower electrodes and "T-shaped protrusions" on top of the upper electrodes [47]. The pressure applied by the ridges of a finger is relayed to the upper electrodes via the T-shaped protrusions and the cavities are deformed. This results in larger capacitances, which are detected by pulsing the lower electrodes. Therefore, this is equivalent to an array of pressure sensors. Because a finger is isolated from the capacitance at each pixel, image acquisition is less susceptible to the finger condition.

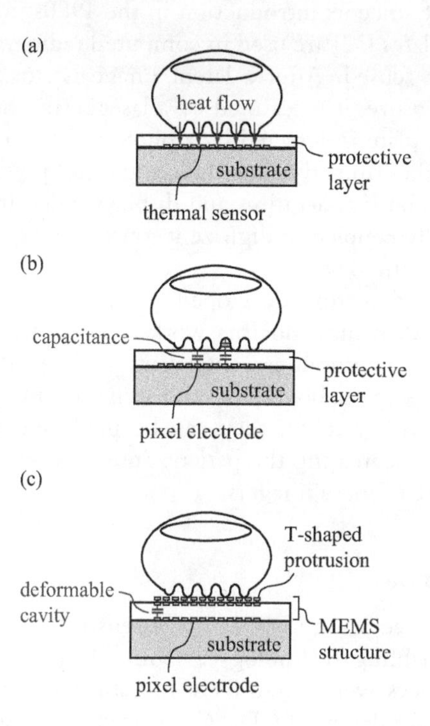

Figure 4.13 Fingerprint sensors based on detecting (a) heat flow, (b) capacitive coupling, and (c) deformation of cavities.

4.4 Flat-panel radiation detectors

In radiology and non-destructive inspection, X-ray photons transmitted through an object are detected by an imager. In nuclear medicine, radioisotopes are introduced into a body, and gamma rays emitted by the decaying nuclei are detected. In both cases, large-area detectors are needed to cover chests, abdomens, breasts, suitcases, etc. Depending on the object to be imaged, an appropriate high voltage is applied to an X-ray tube. Low-energy X-ray photons are filtered to eliminate unwanted exposure. Average X-ray energy in a typical X-ray spectrum is about one-third to one-half of the acceleration voltage applied to an X-ray tube [48].

Various X-ray detectors are available for radiology [49]. X-ray films are still used to detect and record static images. In an X-ray film, phosphor sheets sandwich a photographic plate to convert X-ray to visible photons. Compounds of gadolinium (atomic number 64) are the most common phosphor materials. Exposure to X-rays forms a latent image in the photographic plate. While it is developed, chemical amplification occurs in the photosensitive material. Then, it is fixed to prevent further reactions to light. The recorded images are used for diagnosis and archiving.

Invention of computed tomography (CT) [50] opened the era of digital imaging in 1973. A technology known as computed radiography had been in service clinically since its introduction in the 1980s [51]. Photostimulable storage phosphor plates [52] are used in computed radiography. Unlike X-ray films, they can be reused. After a latent image is stored in the phosphor plate by X-ray exposure, it is scanned by a laser beam. Stimulated emission from the phosphor plate is detected by a photomultiplier tube. Because this scanning process takes time, dynamic imaging is not possible. In contrast, an image intensifier tube [53] acquires and displays video images in real time. A camera is optically coupled to digitize its output images. Image intensifier tubes are bulky vacuum devices.

Flat-panel X-ray detectors developed in the 1990s have changed the way radiologists and medical doctors work at hospitals. Digital images are acquired and displayed without user intervention. These detectors are applied for both static imaging (radiography) and dynamic imaging (fluoroscopy). Unlike image intensifier tubes, they can be made mobile. Much research effort is focused on decreasing the patient dose by improving the detection efficiency and SNR of these imagers.

4.4.1 Early studies

Flat-panel X-ray detectors are MOS image sensors fabricated on large glass substrates. The enabling technology is thin-film semiconductors. In the 1980s, a-Si:H devices were extensively investigated for harvesting energy from sunlight and for driving LCDs. Compared to crystalline counterparts, amorphous semiconductors are suited for large-area applications. They are less vulnerable to radiation damage. It was soon recognized that potential applications of a-Si:H include radiation detectors for medical imaging and high-energy physics experiments.

There are some noteworthy attempts to use a-Si:H layers for radiation detection in the 1980s. In 1985, a-Si solar cells were studied for possible applications in X-ray computed tomography [54]. A p-i-n a-Si:H solar cell was coupled to either a phosphor layer of ZnS or a crystal scintillator $CdWO_4$. In both cases, a good linear relationship was observed between the output current and the incident X-ray photon flux. In 1986, alpha particles from a radioisotope ^{241}Am source were detected by relatively thick (2 μm–15 μm) p-i-n a-Si:H diodes and a stacked n-i-p/metal/p–i–n device for possible applications in high-energy physics experiments [55]. In 1988, β particles from a radioisotope ^{90}Sr source were detected by 27 μm-thick p-i-n a-Si:H diodes [56]. In 1989, a metal/a-Si multilayer was investigated for radiation detection [57]. Metal layers convert X-ray photons to high-energy electrons via Compton scattering and photoelectric effect. The a-Si layers between metal layers convert them to electron–hole pairs.

For imaging, an array of detectors is required. In 1988, a Gd_2O_2S phosphor sheet was coupled to an a-Si:H linear image sensor driven by crystalline Si MOS integrated circuits [58]. Its sensitive length was 256 mm and it had 2,084 pixels. Although the pixel pitch was 125 μm, spatial resolution tested with an 8 keV X-ray beam was 500 μm. Scintillation light spreading inside the phosphor sheet was at least partially responsible for this degradation. In 1989, an a-Si photodiode array with 16 elements was fabricated directly on top of a ceramic scintillator plate of Gd_2O_2S [59]. This is an example of monolithic integration. In 1991, evaporated CsI(Tl) layers were coupled to an a-Si linear image sensor [60]. Evaporated layers have columnar structures. Compared to Gd_2O_2S phosphor sheet, spatial resolution was improved due to confinement of scintillation light. In fact, this approach originated from evaporated CsI(Na) layers used in image intensifiers [61].

Highly sensitive imagers are needed to detect minimum ionizing particles in high-energy physics experiments and to reduce radiation doses in medical imaging. Pixel-level amplification was proposed for radiation imaging in 1989 [62]. Amplifier designs with a-Si TFTs and poly-Si TFTs in a 300 μm× 300 μm pixel were reported in 1990 [63]. Experiments on amplifiers made of high-temperature (solid-phase crystallization) poly-Si TFTs were reported in 1993 [64].

These early studies clearly showed the advantages of a-Si:H layers: large-area fabrication and radiation hardness. Furthermore, they demonstrated some valuable concepts that were pursued later. These include monolithic integration of detector materials on readout electronics, conversion of high-energy radiation to detectable forms, confinement of scintillation light, and pixel-level amplification.

There are two approaches to radiation detection: indirect and direct. Scintillators and photodiodes are used in the first approach. The second approach utilizes semiconductors to convert incident radiation to electron–hole pairs directly. A review paper published in 2008 concluded that indirect imagers outperform direct imagers in standard projection radiography [65]. This view is shared by the authors of a paper published in 2019. According to the authors, "indirect flat-panel detectors are now widely used for medical radiography, fluoroscopy, tomosynthesis, and cone-beam computed tomography" [66]. The two detection principles are described below.

4.4.2 Imagers based on indirect detection

R. A. Street et al. reported the first flat-panel X-ray imagers in 1990. As illustrated in Figure 4.14, each pixel consists of an a-Si photodiode and an a-Si TFT. The imagers had an array of 64 × 40 a-Si photodiodes with pixel sizes ranging from 250 μm to 850 μm. Three metal lines called gate line, data line, and bias line are connected to each pixel as shown. Each gate line is connected to a driver circuit (a shift register with buffers). Each data line is connected to a charge-sensitive amplifier. To convert X-ray to visible photons, a phosphor sheet developed for X-ray films (Gd_2O_2S:Tb) is coupled to the a-Si photodiode array. The first image was a metal screw irradiated by a circular X-ray beam generated at 30 kV [67].

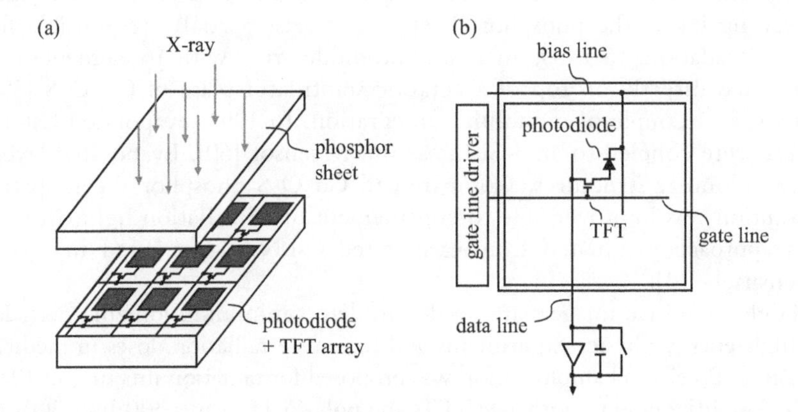

Figure 4.14 Configuration of a flat-panel X-ray imager utilizing an a-Si photodiode array and a phosphor sheet.

Signal size of a MOS image sensor is proportional to the integration time. If the leakage current of a-Si TFTs and the dark current of a-Si photodiodes are low enough, one can accumulate charges at each photodiode for a long time. Using a ceramic scintillator plate coupled to a 64×40 a-Si imager with a pixel pitch of 0.9 mm, gamma-ray imaging was demonstrated with a weak radioisotope source of ^{57}Co emitting 122 keV gamma rays [68].

Another prototype array with 256×240 pixels was developed [69]. Its photograph is shown in Figure 4.15. The pixel pitch is 0.45 mm. The concentric pattern in the photodiode area is due to interference of the microscope illumination light in the top protective coating.

A lens was embedded in the top surface of a black cardboard box. The photodiode array covered by this box worked as a giant digital camera. Sample images taken by this setup are shown in Figure 4.16. The top two images were taken by using one-half of this device. The bottom two are X-ray images taken with this array coupled to a phosphor sheet (Gd_2O_2S:Tb).

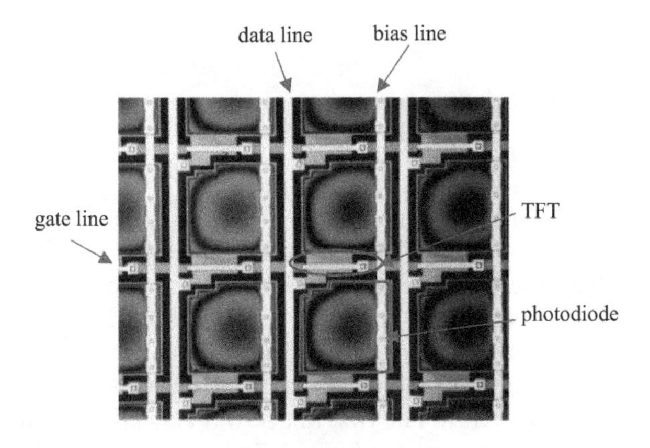

Figure 4.15 Microscope photograph of the pixels in another prototype imager.

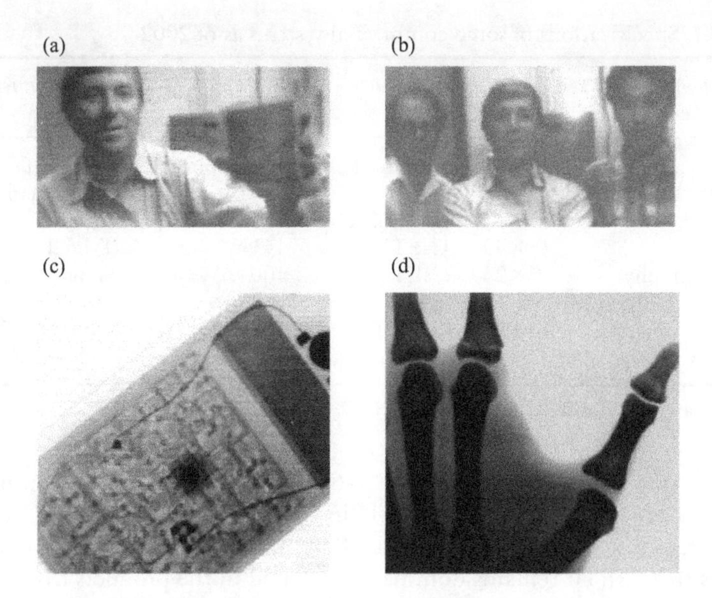

Figure 4.16 Sample images taken by a prototype imager with 256 × 240 pixels. (a) Bob Street holding one of these imagers. (b) Group photo (from left to right: Steve Nelson, Bob Street, Ichiro Fujieda). (c) X-ray image of a calculator. (d) X-ray image of a hand phantom.

Spatial resolution in radiation imaging is determined by many factors in addition to the pixel pitch of an imager. For example, scattered X-ray photons from Compton events are to be rejected because they would blur images. This is usually accomplished by an anti-scatter grid called collimator. This is an array of X-ray absorbers with parallel holes. Obliquely incident photons are absorbed by the collimator. Fine grids are desired for high-resolution imaging. It is reported that MEMS technology can reduce this dimension by one order of magnitude compared to conventional techniques [70]. However, a low fill factor would decrease detection efficiency, leading to higher dose to a patient.

Compton scattering is not the only mechanism that limits spatial resolution. Thick materials with high atomic numbers are required to detect X-rays and gamma rays efficiently. Depth of interactions in a detecting medium is governed by the Lambert–Beer law (see Section 2.1). The point of interaction spreads in space by energy cascade following the initial interaction. Scintillation light can spread inside a scintillator. In this regard, the thickness of a scintillator, the pixel pitch of an a-Si TFT array, and the spectrum of X-ray photons are important design parameters. For example, compared to mammography, X-ray energy for imaging chests is higher. A thicker scintillator is needed to detect these X-ray photons efficiently. Hence, there is a trade-off relationship between spatial resolution and detection efficiency.

Table 4.1 Specifications of some commercial systems as of 2002

Application, typical X-ray tube bias	*Area [cm²]*	*Scintillator*	*Pixel pitch [μm]*	*Manufacturer*
Static imaging (chest) 150 kV	35 × 43	$BaFBr_{1-x}I_x$:Eu	Not applicable	Fuji, Agfa, Eastman Kodak
	43 × 43	Gd_2O_2S:Tb	160	Agfa
	43 × 43	CsI(Tl)	143	Trixell
Mammography 27 kV	19 × 23	CsI(Tl)	100	General electric
Dynamic imaging 60–80 kV	~40 × 40	CsI(Tl)	100	Under development

Source: Carel W E van Eijk, Phys. Med. Biol. 47, R85–R106 (2002).

Dominant scintillators used with a-Si photodiode arrays in commercial systems were Gd_2O_2S:Tb or CsI(Tl) in 2002 [71]. Specifications of some commercial systems are summarized in Table 4.1. At the time of this writing, it appears that CsI(Tl) remains dominant. It is used in the products from Trixell (a joint venture of Thales, Philips Healthcare, and Siemens Healthineers), General Electric, and Canon (Table 4.1) [72].

Studies on the control of scintillation light were reported since the early 1990s. Three examples are noteworthy. First, texturing the surface of a substrate is effective for controlling the columnar structures of CsI(Tl) during evaporation [73]. Second, a similar light-guiding structure filled with a granular phosphor material (Gd_2O_2S:Tb) was reported in 2005 [74]. MEMS technology was applied to fabricate a grid of up to 2 mm tall with 50 μm-thick walls. The surfaces of the grid walls were coated with Al to make them reflective. However, scintillation photons were not well confined because of scattering by the granular material. The fabrication technique of grids might find other applications. Third, front- and back-irradiation geometries were compared in 2019 [75]. Because absorption of photon flux is governed by the Lambert–Beer law, scintillation events occur more frequently near the incident side. Hence, irradiation through an a-Si array results in a better spatial resolution compared to the conventional front-irradiation geometry. In the experiment, a 290 μm-thick Gd_2O_2S:Tb screen and columnar CsI(Tl) layers of thickness up to 1,000 μm were coupled to a commercial a-Si photodiode array with an 85 μm pixel pitch. They also pointed out that the expected lifetime dose for radiography and fluoroscopy detectors is less than 10^2 Gy and that a-Si photodiodes and TFTs retain their characteristics at 10^4 Gy. CMOS active-pixel sensors are expected to be more susceptible to radiation-induced damage. Therefore, pixel-level amplification with poly-Si TFTs might pose a problem in back-irradiation geometry.

Regarding the development of TFT backplanes, a-Si TFT arrays with p-i-n a-Si photodiodes were fabricated on a plastic substrate at low temperatures, and a demonstration of X-ray imaging with a flexible imager was reported in 2009 [76]. In the 2010s, a-IGZO TFTs were replacing a-Si TFTs to take advantage of their higher mobility for applications in LCDs. They were adopted by an X-ray imager in 2012 [77]. Each pixel had an a-IGZO TFT and a p-i-n a-Si:H photodiode. The imager was fabricated on a plastic substrate with a maximum process temperature of 170°C. Combined with a phosphor sheet Gd_2O_2S:Tb, X-ray imaging was demonstrated. In 2016, an organic photodiode array was integrated on an a-IGZO TFT array on a plastic substrate. When coupled to a commercial CsI(Tl) scintillator, a standard lead test chart was imaged with 70 kV X-rays [78].

These studies have demonstrated the feasibility of indirect X-ray imagers based on various combinations of scintillators, photodetectors, TFTs, and substrates. The two approaches described below suggest other possibilities for indirect X-ray imagers.

First, crystalline Si-based CMOS image sensors can be tiled together and coupled to a scintillator. In the paper published in 2009 [79], a CMOS active-pixel sensor (APS) with a very large pixel pitch of 96 μm was developed. Four of them were placed side by side with minimum gaps between them. To make room for driver circuits at the center, the width of the edge pixels in each sensor was reduced to 60 μm. The tiled sensor had 1,024 × 512 pixels and an active area of 49.1 mm × 98.3 mm. A fiber optic plate was used to couple a scintillator to this tiled sensor. Low-noise X-ray imaging was demonstrated for X-ray tube bias at 50 kV. This tiled sensor approach might provide a good solution for small to medium-area imaging applications. There is another study of using a CMOS APS with large pixels for mammography [80]. A commercial CMOS APS with a 50 μm pixel pitch was coupled to a 165 μm-thick structured CsI(Tl) layer. The sensor had active pixels of 1,032 × 1,032 for an active area of 51.6 mm × 51.6 mm.

Second, the use of indirect X-ray imagers for single-photon emission computed tomography (SPECT) was reported in 2020 [81]. In this proof-of-concept experiment, a parallel-hole collimator was placed on a commercial X-ray detector using CsI(Tl) from Trixell. Projection data were acquired with 141 keV gamma rays from a ^{99m}Tc source placed nearby. They concluded that this configuration would give a cost-effective alternative to purchasing a conventional SPECT. Nevertheless, it is interesting to see if the optimum design should improve the performance of indirect imagers in nuclear medicine.

4.4.3 Imagers based on direct detection

An X-ray imager based on direct detection is an example of monolithic integration of a conversion material on a readout circuit. As shown schematically in Figure 4.17, a continuous semiconductor layer and a top electrode are formed on a TFT array with pixel electrodes. A high voltage is usually applied across the semiconductor layer. Blocking layers are needed to prevent charge injection from the electrodes. When X-ray photons are

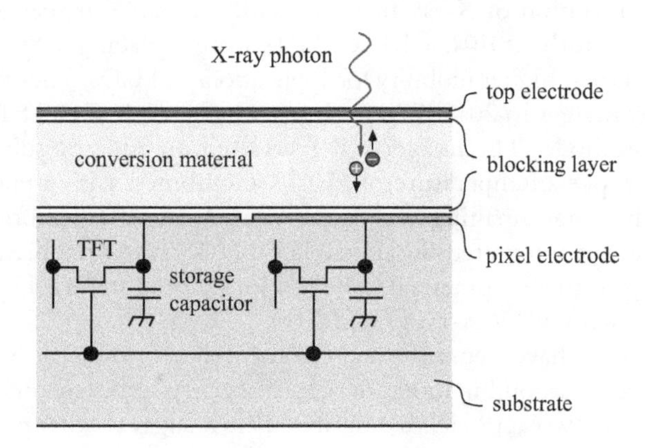

Figure 4.17 Cross section of the pixel structure of a direct conversion imager.

absorbed, electron–hole pairs are created. They drift along the electric field and signal charges are induced on pixel electrodes. They are stored by the capacitor at each pixel and are read out by turning on their TFT. In contrast to scintillation photons in a phosphor without any confinement structures, the lateral spreading of charge carriers in the biased semiconductor layer is limited by the presence of electric field. Therefore, higher spatial resolution is expected for direct detection in principle.

Amorphous selenium (a-Se) had been used in xerography until inexpensive organic photoconductors replaced it in the late 1980s [82]. In 1995, X-ray imaging was demonstrated with an a-Se layer formed on an a-Si TFT array [83]. An X-ray image of a hand phantom was read out by a 1,280 × 1,536 a-Si TFT array driven by external crystalline Si-integrated circuits. The pixel size was 139 μm× 139 μm. The thickness of the a-Se layer was 300 μm and the electric field was 4.2 V/μm. A design consideration on an imager based on an a-Se layer and a CdSe TFT array was also presented at the same conference in 1995 [84].

The electron–hole pair creation energy (ε value) of a-Se depends on the electric field. It varies from 30 eV to 70 eV at 10 V/μm [85]. For example, a 50 keV X-ray photon would be converted to 1,000 electrons if it were 50 eV.

To increase the signal charge, configurations based on avalanche multiplication inside a-Se were proposed. The structure shown in Figure 4.18a was proposed in 2005 [86]. It adopts an avalanche-mode a-Se photoconductive layer developed for a high-sensitivity camera tube [87]. A structured CsI(Tl) is coupled to this multilayered photosensor. Scintillation photons generate electrons and holes in the a-Se layer. Holes drift toward the bottom electrode. Under a sufficiently high electric field (>70 V/μm), impact ionization takes place. Pixel electrodes of a TFT array read out these amplified charges. In the configuration proposed in 2008, an a-Se layer is divided into two regions by a mesh electrode [88]. As illustrated in Figure 4.18b, X-ray photons are

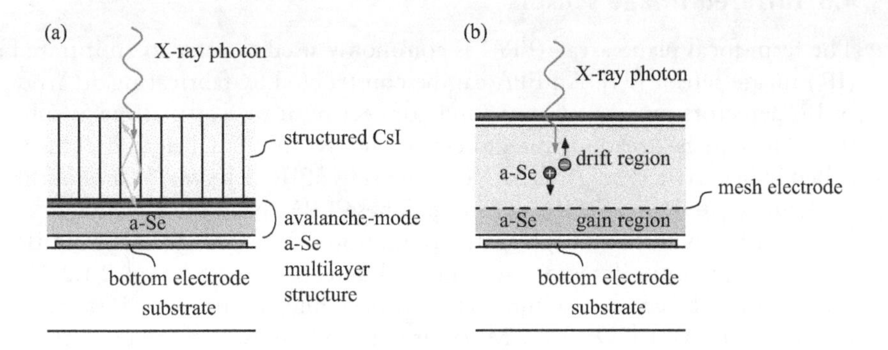

Figure 4.18 Configurations for utilizing avalanche multiplication in an a-Se layer: (a) a structured scintillator is stacked on the a-Se multilayer structure and (b) an a-Se layer is divided into drift and avalanche regions by a mesh electrode.

converted to electrons and holes in the first region, and the holes drift toward the bottom electrode. The majority of these holes enter the second region. Under a sufficiently high electric field, the number of holes increases by impact ionization. In both configurations, dark current must be suppressed because these background charges are also multiplied. If this is the case, there is no merit of internal gain for SNR. Demonstration of X-ray imaging with prototypes would prove feasibility of these concepts.

It is interesting to note that the idea of dividing conversion and multiplication processes is analogous to a multiwire proportional chamber (MWPC) [89]. In this gas-filled position-sensitive radiation detector used in high-energy physics experiments, incident radiation ionizes gas molecules. Electrons drift toward small-diameter wires. The large electric field in the vicinity of the wire induces impact ionization. A solid-state device equivalent to an MWPC might be an interesting topic to investigate in future.

The atomic number of Se is 34. Because it is relatively low, high-energy X-ray photons are not absorbed efficiently. This is the reason why direct conversion imagers utilizing a-Se are more suited for mammography rather than chest imaging [90]. Semiconductors HgI_2 and PbI_2 contain heavy metals and their ε values are about one-tenth of that of an a-Se layer under 10 V/μm. They can be evaporated over a large area. These features make them attractive candidates for replacing a-Se. However, detailed studies on these materials revealed that each material had its own problems such as incomplete charge collection, large dark current, and image lag [91]. More recent candidates are perovskites containing heavy elements such as cesium lead bromide $CsPbBr_3$ [92,93]. It would be interesting to see if these materials can fulfill their potential in direct flat-panel X-ray imagers.

4.5 Infrared image sensors

The term focal plane array (FPA) is commonly used to refer to an infrared (IR) image sensor [94]. An FPA can be constructed by fabricating an array of IR detectors on one substrate and connecting it to a crystalline Si substrate on which a readout integrated circuit (ROIC) is fabricated. This is called hybrid structure. Alternatively, an array of IR detectors is monolithically integrated on a ROIC substrate. These FPAs have the advantages of low fabrication cost and high spatial resolution. There are two types of IR detectors: quantum detectors and thermal detectors (see Section 2.1.3). In general, FPAs based on quantum detectors have higher sensitivity. Regarding ROICs for FPAs, CCDs and CMOS circuits on Si substrates are used. As in the case of visible light imaging, CMOS ROICs for IR imaging [95] have the advantages of versatile operations, low power consumption, and low fabrication cost.

4.5.1 Hybrid structures

Flip-chip bonding technology uses indium bumps for connecting a detector substrate to a ROIC Si substrate as illustrated in Figure 4.19a. Indium forms a good bond at room temperature and endures thermal cycling between 300 K and the usual operating temperature of 77 K [96]. The FPAs can be illuminated from both sides. Illumination through the detector substrate is preferred because the ROIC Si substrate has metallization and other opaque regions.

The loophole (or via-hole) interconnection technique shown in Figure 4.19b has advantages in mechanical strength and thermal stability. First, a detector substrate such as p-type HgCdTe is glued to a ROIC Si substrate. Via holes are drilled on the detector substrate by ion milling. Then, p-n junctions are created in either ion implantation or diffusion process. Metallization connects the two metal pads. Finally, the surface is passivated. Obviously, the regions occupied by via holes are not sensitive to light. Therefore, the loophole technique has a problem of reduced fill factor [97].

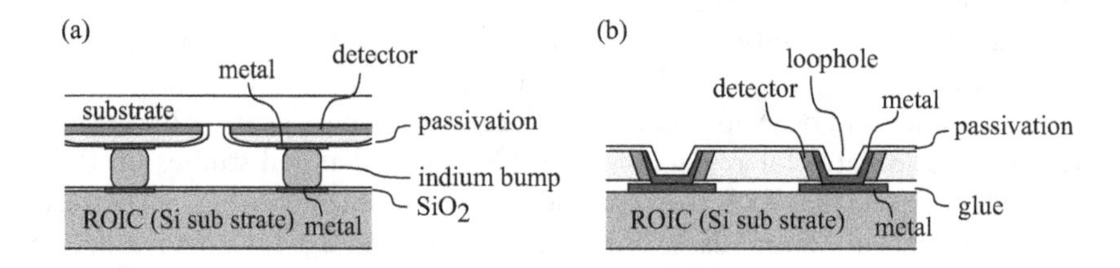

Figure 4.19 Hybrid structures based on (a) indium bumps and (b) loopholes.

4.5.2 Monolithic integration

A fully monolithic indium antimonide (InSb) infrared CCD array was reported in 1980 [98]. First, an array of a p–n junction and a MOS structure was formed on an InSb wafer. Then, a CCD was fabricated on an InSb wafer. It was a linear image sensor with only 20 pixels. Rather than detecting IR photons in a CCD itself, a separate array of photodetectors allowed one to control integration time independent of the clock timing for the CCD operation. However, narrow-band semiconductor CCDs had a problem of trapping of charge carriers that decreases transfer efficiency. Therefore, a large format array was not practical.

Developments of uncooled two-dimensional FPAs started in the late 1970s for military and consumer applications. In 1992, an array of microbolometers (336 × 230 pixels) was monolithically integrated on a crystalline Si ROIC [99]. The pixel size was 2 mils × 2 mils (1 mil = 0.001 inch = 25.4 μm) with a fill factor of 0.48. Major efforts since then have advanced the performance of uncooled FPAs based on microbolometers. For example, initial microbolometers had a single-layer structure as illustrated in Figure 4.20 [100]. In the two-level structure shown in Figure 4.20b, a separate IR absorber is connected to a serpentine resistor via posts [101]. Now, the absorber covers the supporting legs and metal lines like an umbrella. Thus, the fill factor increases. The pixel size of a microbolometer array was reduced to 20 μm around 2010 [102].

Furthermore, designs of IR absorbers added new functions. For example, an array with polarization-selective absorbers was developed by forming grating structures with tapered sidewalls [103]. In addition to pixel designs, advances in CMOS readout circuits continue [104]. A steady market growth in automobile and security applications is expected for low-cost microbolometer FPAs [105].

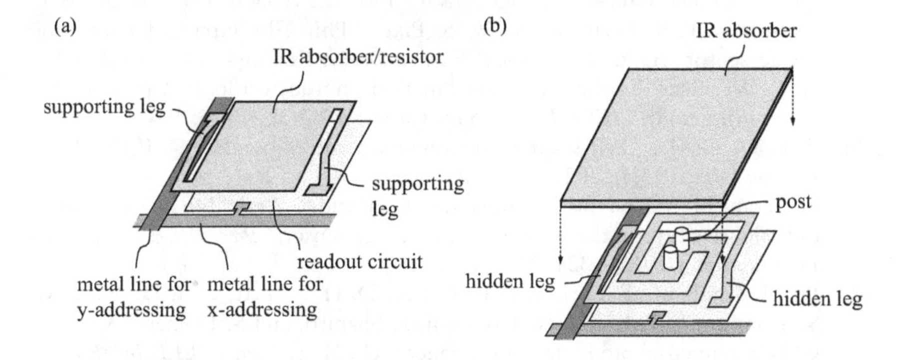

Figure 4.20 Structures of microbolometers. (a) A single-layer design and (b) a double-level design with a separate IR absorber connected to a serpentine resistor via posts.

For detecting near-infrared photons at room temperature, thin-film quantum dots (QDs) are investigated for monolithic integration with CMOS readout electronics. Using lead sulfide (PbS) QDs, external quantum efficiency (EQE) above 20% at a wavelength of 1.44 μm was reported. The peak absorption wavelength of these QD films shifted from 1,440 nm to 980 nm when the diameter of QDs varied from 5.5 nm to 3.4 nm [106]. When the bias across the QD layer was between −0.5 and 0 V, the EQE at a wavelength of 940 nm was nearly zero and it was above 30% beyond 1 V. This feature of QD sensors added a global shutter feature to a camera, which allowed image capture at a single point of time [107]. A review paper published in 2022 reports that some QD image sensors have EQE larger than 60% in the visible range [108].

References

1. G. P. Weckler, "Operation of p-n junction photodetectors in a photon flux integrating mode," *J. Solid-State Circuits* SC-2(3), 65–73 (1967).
2. W. S. Boyle and G. E. Smith, "B.S.T.J. brief: Charge coupled semiconductor devices," *Bell Syst. Tech. J.* 49, 587–593 (1970).
3. E. R. Fossum, "Active pixel sensors: Are CCD's dinosaurs?" *Proc. SPIE* 1900, 2–14 (1993).
4. I. Fujieda, G. Cho, S. N. Kaplan, V. Perez-Mendez, S. Qureshi, R. A. Street, "Applications of a-Si:H radiation detectors," *J. Non-Cryst. Solids* 115, 174–176 (1989).
5. M. Kimata, "Uncooled infrared focal plane arrays," *IEEJ Trans. Elec. Electron. Eng.* 13, 4–12 (2018).
6. M. A. Schuster and G. Strull, "A monolithic mosaic of photon sensors for solid-state imaging applications," *IEEE Trans. Electron Devices* ED-13(12), 907–912 (1966).
7. P. K. Weimer, G. Sadasiv, J. E. Meyer, L. Meray-Horvath and W. S. Pike, "A self-scanned solid-state image sensor," *Proc. IEEE* 55(9), 1591–1602 (1967).
8. G. Sadasiv, P. K. Weimer, and W. S. Pike, "Thin-film circuits for scanning image-sensor arrays," *IEEE Trans. Electron Devices* ED-15(4), 215–219 (1968).
9. G. P. Weckler, "Operation of p-n junction photodetectors in a photon flux integrating mode," *IEEE J. Solid-State Circuits* SC-2(3), 65–73, 1967.
10. P. J. W. Noble, "Self-scanned silicon image detector arrays," *IEEE Trans. Electron Devices* 15(4), 202–209 (1968).
11. N. Teranishi, A. Kohono, Y. Ishihara, E. Oda, and K. Arai, "No image lag photodiode structure in the interline CCD image sensor," *Proc. IEEE Int. Electron Devices Meet. (IEDM)*, 324–327 (1982).
12. R. M. Guidashf, T.-H. Lee, P. P. K. Lee, D. H. Sackett, C. I. Drowley, M. S. Swenson, L. Arbaugh, R. Hollstein, F. Shapiro, and S. Domer, "A 0.6 μm CMOS pinned photodiode color imager technology," *Proc. IEEE Int. Electron Devices Meet. (IEDM)*, 927–929 (1997).
13. K. Komiya, M. Kanzaki, and T. Yamashita, "A 2048-element contact-type linear image sensor for facsimile," *Proc. IEEE Int. Electron Devices Meet. (IEDM)*, 309–312 (1981).

14. S. Kaneko, M. Sakamoto, F. Okumura, T. Itano, H. Kataniwa, Y. Kajiwara, M. Kanamori, M. Yasumoto, T. Saito, and T. Ohkubo, "Amorphous Si:H contact linear image sensor with Si_3N_4 blocking layer," *Proc. IEEE Int. Electron Devices Meet. (IEDM)*, 328–331 (1982).
15. M. Kunii, K. Hasegawa, H. Oka, Y. Nakazawa, T. Takeshita, and H. Kurihara, "Performance of a high-resolution contact-type linear image sensor with a-Si:H/a-SiC:H heterojunction photodiodes," *IEEE Trans. Electron Devices* 36(12), 2877–2882 (1989).
16. S. Morozumi, H. Kurihara, T. Takeshita, H. Oka, and K. Hasegawa, "Completely integrated contact-type linear image sensor," *Trans. Electron Devices* 32(8), 1546–1550 (1985).
17. T. Kaneko, Y. Hosokawa, M. Tadauchi, Y. Kita, and H. Andoh, "400 dpi integrated contact type linear image sensors with poly-Si TFT's analog readout circuits and dynamic shift registers," *IEEE Trans. Electron Devices* 38(5), 1086–1093 (1991).
18. I. Fujieda, F. Okumura, K. Sera, H. Asada, and H. Sekine, "Self-referenced poly-Si TFT amplifier readout for a linear image sensor," *Proc. IEEE Int. Electron Devices Meet. (IEDM)*, 587–590 (1993).
19. J. D. Plummer and J. D. Meindl, "MOS electronics for a portable reading aid for the blind," *IEEE J. Solid-State Circuits* SC-7(2), 111–119 (1972).
20. L. K. J. Vandamme and F. N. Hooge, "What do we certainly know about 1/f noise in MOSTs?" *IEEE Trans. Electron Devices* 55(11), 3070–3085 (2008).
21. A. M. Fowler, R. G. Probst, J. P. Britt, R. R. Joyce, and F. C. Gillett, "Evaluation of an indium antimonide hybrid focal plane array for ground-based infrared astronomy," *Opt. Eng.* 26, 232–240 (1987).
22. E. R. Fossum "Active pixel sensors: Are CCDs dinosaurs?" *Proc. SPIE* 1900, 2–14 (1993).
23. M. H. White, D. R. Lampe, F. C. Blaha and I. A. Mack, "Characterization of surface channel CCD image arrays at low light levels," *IEEE J. Solid-State Circuits* 9(1), 1–12 (1974).
24. P. Martin-Gonthier and P. Magnan, "RTS noise impact in CMOS image sensors readout circuit," *2009 16th IEEE International Conference on Electronics, Circuits and Systems – (ICECS 2009)*, 2009, pp. 928–931.
25. E. R. Fossum, "CMOS image sensors: Electronic camera-on-a-chip," *IEEE Trans. Electron Devices* 44(10), 1689–1698 (1997).
26. H. Takahashi, M. Kinoshita, K. Morita, T. Shirai, T. Sato, T. Kimura, H. Yuzurihara, and S. Inoue, "A 3.9 μm pixel pitch VGA format 10 b digital image sensor with 1.5-transistor/pixel," *2004 IEEE International Solid-State Circuits Conference*, 2004, pp. 108–516.
27. K-B. Cho, C. Lee, S. Eikeda, A. Baum, J. Jiang, C. Xu, X. Fan, and R. Kauffman, "A 1/2.5 inch 8.1Mpixel CMOS image sensor for digital cameras," *2007 IEEE International Solid-State Circuits Conference*, 2007, pp. 508–618.
28. K. Itonaga, K. Mizuta, T. Kataoka, M. Yanagita, H. Ikeda, H. Ishiwata, Y. Tanaka, T. Wakano, Y. Matoba, T. Oishi, R. Yamamoto, S. Arakawa, J. Komachi, M. Katsumata, S. Watanabe, S. Saito, T. Haruta, S. Matsumoto, K. Ohno, T. Ezaki, T. Nagano, and T. Hirayama, "Extremely-low-noise CMOS image sensor with high saturation capacity," *Proc. IEEE Int. Electron Devices Mee. (IEDM)*, 8.1.1–8.1.8 (2011).

29. E. R. Fossum and D. B. Hondongwa, "A review of the pinned photodiode for CCD and CMOS image sensors," *IEEE J. Electron Devices Soc.* 2(3), 33–43 (2014).
30. R. Fontaine, "The evolution of pixel structures for consumer-grade image sensors," *IEEE Trans. Semicond. Manuf.* 26(1), 11–16 (2013).
31. S. Iwabuchi, Y. Maruyama, Y. Ohgishi, M. Muramatsu, N. Karasawa, and T. Hirayama, "A back-illuminated high-sensitivity small-pixel color CMOS image sensor with flexible layout of metal wiring," *IEEE Int. Solid-State Circuits Conf. – Dig. Tech. Pap.* 1171–1178 (2006).
32. S. R. Shortes, W. W. Chan, W. C. Rhines, J. B. Barton, and D. R. Collins, "Characteristics of thinned backside-illuminated charge-coupled device imagers," *Appl. Phys. Lett.* 24, 565–567 (1974).
33. Y. Huo, C. C. Fesenmaier, and P. B. Catrysse, "Microlens performance limits in sub-2μm pixel CMOS image sensors," *Opt. Express* 18(6), 5861 (2010).
34. J. P. Lu, K. Van Schuylenbergh, J. Ho, Y. Wang, J. B. Boyce, and R. A. Street, "Flat panel imagers with pixel level amplifiers based on polycrystalline silicon thin-film transistor technology," *Appl. Phys. Lett.* 80, 4656–4658 (2002).
35. R. A. Street, J. Graham, Z. D. Popovic, A. Hor, S. Ready, and J. Ho, "Image sensors combining an organic photoconductor with a-Si:H matrix addressing," *J. Non-Cryst. Solids* 299–302, 1240–1244 (2002).
36. T. N. Ng, W. S. Wong, M. L. Chabinyc, S. Sambandan, and R. A. Street, "Flexible image sensor array with bulk heterojunction organic photodiode," *Appl. Phys. Lett.* 92, 213303 (2008).
37. V. Pejovi, E. Georgitzikis, J. Lee, I. Lieberman, D. Cheyns, P. Heremans, and P. E. Malinowski, "Infrared colloidal quantum dot image sensors," *IEEE Trans. Electron Devices* 69(6), 2840–2850 (2022).
38. G. E. Smith, "Nobel lecture: The invention and early history of the CCD," *Rev. Mod. Phys.* 82, 2307 (2010).
39. W. Kosonocky and J. E. Carnes, "Charge-coupled digital circuits," *IEEE J. Solid-State Circuits* 6(5), 314–322 (1971).
40. R. H. Walden, R. H. Krambeck, R. J. Strain, J. Mckenna, N. L. Schryer, and G. E. Smith, "B.S.T.J. brief: The buried channel charge coupled device," *Bell Syst. Tech. J.* 51(7), 1635–1640 (1972).
41. W. H. Kent, "Charge distribution in buried-channel charge-coupled devices," *Bell Syst. Tech. J.* 52(6), 1009–1024 (1973).
42. A. M. Mohsen, R. Bower, T. C. McGill, and T. Zimmerman, "Overlapping-gate buried-channel charge-coupled devices," *Electron. Lett.* 9(17), 396–398 (1973).
43. N. J. Harrick, "Fingerprinting via total internal reflection," *Phillips Tech. Rev.* 24, 271–274 (1962/63).
44. Y. Koda, T. Higuchi and A. K. Jain, "Advances in capturing child fingerprints: A high resolution CMOS image sensor with SLDR method," *2016 International Conference of the Biometrics Special Interest Group (BIOSIG)*, 2016, pp. 1–4.
45. H. Akkerman, B. Peeters, D. Tordera, A. van Breemen, S. Shanmugam, P. Malinowski, J. Maas, J. de Riet, R. Verbeek, T. Bel, G. de Haas, D. Hermes, S. K. Del, A. J. Kronemeijer, and G. Gelinck, "Large-area optical fingerprint sensors for next generation smartphones," *SID Symp. Dig. Tech. Pap.* 50, 1000–1003 (2019).

46. R. Hashido, A. Suzuki, A. Iwata, T. Okamoto, Y. Satoh, and M. Inoue, "A capacitive fingerprint sensor chip using low-temperature poly-Si TFTs on a glass substrate and a novel and unique sensing method," *IEEE J. Solid-State Circuits* 38(2), 274–280 (2003).
47. N. Sato, S. Shigematsu, H. Morimura, M. Yano, K. Kudou, T. Kamei, and K. Machida, "Novel surface structure and its fabrication process for MEMS fingerprint sensor," *IEEE Trans. Electron Devices* 52(5), 1026–103 (2005).
48. J. Anthony Seibert, "X-ray imaging physics for nuclear medicine technologists. Part 1: Basic principles of X-ray production," *J. Nucl. Med.* 32(3), 139–147 (2004).
49. M. J. Yaffe and J. A. Rowlands, "X-ray detectors for digital radiography," *Phys. Med. Biol.* 42, 1–39 (1997).
50. G. N. Hounsfield, "Computerized transverse axial scanning (tomography): Part 1. Description of system," *Br. J. Radiol.* 46, 1016–1022 (1973).
51. J. Anthony Seibert and J. M. Boone, "X-ray imaging physics for nuclear medicine technologists. Part 2: X-ray interactions and image formation," *J. Nucl. Med.* 33(1), 3–18 (2005).
52. J. Miyahara, K. Takahashi, Y. Amemiya, N. Kamiya, and Y. Satow, "A new type of X-ray area detector utilizing laser stimulated luminescence," *Nucl. Instr. and Meth. A* 246, 572–578 (1986).
53. W. J. Oosterkamp, "Image intensifier tubes," *Acta Radiologica* 41(sup116), 495–502 (1954).
54. W. Guang-Pu, H. Okamoto, and Y. Hamakawa, "Amorphous-silicon photovoltaic X-ray sensor," *Jpn. J. Appl. Phys.* 24, 1105–1106 (1985).
55. S. N. Kaplan, J. Morel, V. Perez-Mendez, and R. A. Street, "Detection of charged particles in amorphous silicon layers," *IEEE Trans. Nucl. Sci.* NS-33, 351–354 (1986).
56. V. Perez-Mendez, S. N. Kaplan, G. Cho, I. Fujieda, S. Qureshi, W. Ward, and R. A. Street, "Hydrogenated amorphous silicon pixel detectors for minimum ionizing particles," *Nucl. Instr. Meth.* A273, 127–134 (1988).
57. Y. Naruse and T. Hatayama, "Metal/amorphous silicon multilayer radiation detectors," *IEEE Trans. Nucl. Sci.* NS-36, 1347–1352 (1989).
58. K. Mochiki, K. Hasegawa, and S. Namatame, "Amorphous silicon position-sensitive detector," *Nucl. Instr. Meth.* A273, 640–644 (1988).
59. H. Ito, S. Matsubara, T. Takahashi, T. Shimada, and H. Takeuchi, "Integrated radiation detectors with a-Si photodiodes on ceramic scintillators," *Jpn. J. Appl. Phys.*, L1476–L1479 (1989).
60. I. Fujieda, G. Cho, J. Drewery, T. Gee, T. Jing, S. N. Kaplan, V. Perez-Mendez, D. Wildermuth, "X-ray and charged particle detection with CsI(Tl) layer coupled to a Si:H photodiode layers," *IEEE Trans. Nucl. Sci.* 38(2), 255–262 (1991).
61. A. L. N. Stevels and A. D. M. Schramade Pauw, "Vapour-deposited CsI:Na layers, I. morphologic and crystallographic properties," *Philips Res. Repts.* 29, 340–352 (1974).
62. V. Perez-Mendez, G. Cho, I. Fujieda, S. N. Kaplan, S. Qureshi and R. A. Street, "Proposed thin film electronics for a-Si:H PIXEL detectors," Lawrence Berkeley Laboratory, University of California, Berkeley, LBL-26254 (1989).
63. G. Cho, M. Conti, J. S. Drewery, I. Fujieda, S. N. Kaplan, V. Perez-Mendez, S. Qureshi, and R. A. Street, "Assessment of TFT amplifiers for a-Si:H PIXEL particle detectors," *IEEE Trans. Nucl. Sci.* 37(3), 1142–1148 (1990).

64. G. Cho, J. S. Drewery, W. S. Hong, T. Jing, S. N. Kaplan, H. Lee, A. Mireshghi, V. Perez-Mendez, and D. Wildermuth, "Signal readout in a-Si:H pixel detectors," *IEEE Trans. Nucl. Sci.* 40(4), 323–327 (1993).
65. A. R. Cowen, S. M. Kengyelics, and A. G. Davies, "Solid-state, flat-panel, digital radiography detectors and their physical imaging characteristics," *Clin. Radiol.* 63, 487–498 (2008).
66. A. Howansky, A. Mishchenko, A. R. Lubinsky, and W. Zhao, "Comparison of CsI:Tl and Gd_2O_2S:Tb indirect flat panel detector x-ray imaging performance in front- and back-irradiation geometries," *Med. Phys.* 46(11), 4857–4868 (2019).
67. R. A. Street, S. Nelson, L. E. Antonuk, and V. Perez Mendez, "Amorphous silicon sensor arrays for radiation imaging," *MRS Symp. Proc.* 192, 441–452 (1990).
68. I. Fujieda, S. Nelson, P. Nylen, R. A. Street and R. L. Weisfield, "Two operation modes of 2D a-Si sensor arrays for radiation imaging," *J. Non-Crystalline Solids* 137&138, 1321–1324 (1991).
69. I. Fujieda, S. Nelson, R. A. Street, and R. L. Weisfield, "Radiation imaging with 2D a-Si sensor arrays," *IEEE Trans. Nucl. Sci.* 39(4), 1056–1062 (1992).
70. V. Lehmanna and S. RoÈnnebeck, "MEMS techniques applied to the fabrication of anti-scatter grids for X-ray imaging," *Sens. Actuator A Phys.* 95(2–3), 202–207 (2002).
71. C. W. E. van Eijk, "Inorganic scintillators in medical imaging," *Phys. Med. Biol.* 47, R85–R106 (2002).
72. https://www.trixell.com/technology, https://www.gehealthcare.com/products/radiography, https://etd.canon/en/product/category/fpd/index.html (accessed on Sep. 23, 2022).
73. T. Jing, G. Cho, J. Drewery, I. Fujieda, S. N. Kaplan, A. Mireshghi, V. Perez-Mendez, D. Wildermuth, "Enhanced columnar structure in CsI layer by substrate patterning," *IEEE Trans. Nucl. Sci.* 39(5), 1195–1198 (1992).
74. A. Sawant, L. E. Antonuk, Y. El-Mohri, Y. Li, Z. Su, Y. Wang, J. Yamamoto, Q. Zhao, H. Du, J. Daniel, and R. Street, "Segmented phosphors: MEMS-based high quantum efficiency detectors for megavoltage x-ray imaging," *Med. Phys.* 32(2), 553–565 (2005).
75. A. Howansky, A. Mishchenko, A. R. Lubinsky, and W. Zhao, "Comparison of CsI:Tl and Gd_2O_2S:Tb indirect flat panel detector x-ray imaging performance in front- and back-irradiation geometries," *Med. Phys.* 46(11), 4857–4868 (2019).
76. R. A. Street, W. S. Wong, T. Ng, and R. Lujan, "Amorphous silicon thin film transistor image sensors," *Philos. Mag.* 89(28–30), 2687–2697 (2009).
77. R. A. Lujan and R. A. Street, "Flexible X-ray detector array fabricated with oxide thin-film transistors," *IEEE Electron Device Lett.* 33(5), 688–690 (2012).
78. G. H. Gelinck, A. Kumar, D. Moet, J-L. P. J. van der Steen, A. J. J. M. van Breemen, S. Shanmugam, A. Langen, J. Gilot, P. Groen, R. Andriessen, M. Simon, W. Ruetten, A. U. Douglas, R. Raaijmakers, P. E. Malinowski, and K. Myny, "X-Ray detector-on-plastic with high sensitivity using low cost, solution-processed organic photodiodes," *IEEE Trans. Electron Devices* 63(1), 197–204 (2016).
79. M. Farrier, T. Graeve, Achterkirchen, G. P. Weckler, and A. Mrozack, "Very large area CMOS active-pixel sensor for digital radiography," *IEEE Trans. Electron Devices* 56, 2623–2631 (2009).

80. M. Endrizzi, P. Oliva, B. Golosio, and P. Delogu, "CMOS APS detector characterization for quantitative X-ray imaging," *Nucl. Instrum. Methods Phys. Res. A* 703, 26–32 (2013).
81. M. M. A. Dietze, W. J. C. Koppert, R. van Rooij, and H. W. A. M. de Jong, "Nuclear imaging with an x-ray flat panel detector: A proof-of-concept study," *Med. Phys.* 47(8), 3363–3368 (2020).
82. S. Kasap, J. B. Frey, G. Belev, O. Tousignant, H. Mani, L. Laperriere, A. Reznik, and J. A. Rowlands, "Amorphous selenium and its alloys from early xeroradiography to high resolution X-ray image detectors and ultrasensitive imaging tubes," *Phys. Stat. Sol. (B)* 246, 1794–1805 (2009).
83. D. L. Y. Lee, L. K. Cheung, and L. S. Jeromin, "New digital detector for projection radiography," *Proc. SPIE* 2432, 237–249 (1995).
84. W. Zhao, J. A. Rowlands, S. Germann, D. F. Waechter, and Z. Huang, "Digital radiology using self-scanned readout of amorphous selenium: Design considerations for mammography," *Proc. SPIE* 2432, 250–251 (1995).
85. J. L. Donovan, "X-ray sensitivity of selenium," *J. Appl. Phys.* 50, 6500–6504 (1979).
86. W. Zhao, D. Li, A. Reznik, B. J. M. Lui, D. C. Hunt, J. A. Rowlands, Y. Ohkawa, and K. Tanioka, "Indirect flat-panel detector with avalanche gain: Fundamental feasibility investigation for SHARP-AMFPI scintillator HARP active matrix flat panel imager," *Med. Phys.* 32, 2954–2966 (2005).
87. K. Tanioka, J. Yamazaki, K. Shidara, K. Taketoshi, T. Kawamura, T. Hirai, and Y. Takasaki, "An avalanche-mode amorphous Selenium photoconductive layer for use as a camera tube target," *IEEE Electron Device Lett.* 8(9), 392–394 (1987).
88. M. M. Wronski and J. A. Rowlands, "Direct-conversion flat-panel imager with avalanche gain: Feasibility investigation for HARP-AMFPI," *Med. Phys.* 35(12), 5207–5218 (2008).
89. G. Charpak and F. Sauli, "Multiwire proportional chambers and drift chambers," *Nucl. Instrum. Methods* 162, 405–428 (1979).
90. S. Kasap, J. B. Frey, G. Belev, O. Tousignant, H. Mani, J. Greenspan, L. Laperriere, O. Bubon, A. Reznik, G. De Crescenzo, K. S. Karim, and J. A. Rowlands, "Amorphous and polycrystalline photoconductors for direct conversion flat panel x-ray image sensors," *Sensors* 11, 5112–5157 (2011).
91. R. A. Street, S. E. Ready, K. Van Schuylenbergh, J. Ho, J. B. Boyce, and P. Nylen, "Comparison of PbI_2 and HgI_2 for direct detection active matrix x-ray image sensors," *J. Appl. Phys.* 91, 3345–3355 (2002).
92. H. Zhang, F. Wang, Y. Lu, Q. Sun, Y. Xu, B-B. Zhang, W. Jie, and M. G. Kanatzidis, "High-sensitivity X-ray detectors based on solution-grown caesium lead bromide single crystals," *J. Mater. Chem. C* 8, 1248–1256 (2020).
93. Y. Zhou, L. Zhao, Z. Ni, S. Xu, J. Zhao, X. Xiao, and J. Huang, "Heterojunction structures for reduced noise in large-area and sensitive perovskite x-ray detectors," *Sci. Adv.* 7(36), eabg6716 (2021).
94. A. Rogalski, "Progress in focal plane array technologies," *Prog. Quantum. Electron.* 36(2–3), 342–473 (2012).
95. C-C. Hsieh, C-Y. Wu, F-W. Jih, and T-P. Sun, "Focal-plane-arrays and CMOS readout techniques of infrared imaging systems," *IEEE Trans. Circuits Syst. Video Technol.* 7(4), 594–605 (1997).
96. J. Jiang, S. Tsao, T. O'Sullivan, M. Razeghi, and G. J. Brown, "Fabrication of indium bumps for hybrid infrared focal plane array applications," *Infrared Phys. Technol.* 45(2), 143–151 (2004).

97. D. A. Scribner, M. R. Kruer, and J. M. Killiany, "Infrared focal plane array technology," *Proc. IEEE* 79(1), 66–85 (1991).
98. R. D. Thom, T. L. Koch, J. D. Langan, and W. J. Parrish, "A fully monolithic InSb infrared CCD array," *IEEE Trans. Electron Devices* 27(1), 160–170 (1980).
99. R. A. Wood, C. J. Han, and P. W. Kruse, "Integrated uncooled infrared detector imaging arrays," *Technical Digest IEEE Solid-State Sensor and Actuator Workshop*, 1992, pp. 132–135.
100. R. A. Wood, "High-performance infrared thermal imaging with monolithic silicon focal planes operating at room temperature," *Proc. IEEE Int. Electron Devices Meet. (IEDM)*, 175–177 (1993).
101. H-K. Lee, J-B. Yoon, and E. Yoon, "A high fill-factor IR bolometer using multi-level electrothermal structures," *Proc. IEEE Int. Electron Devices Meet. (IEDM)*, 463–466 (1998).
102. L. Yu, Y. Guo, H. Zhu, M. Luo, P. Han, and X. Ji, "Low-cost microbolometer type infrared detectors," *Micromachines* 11(9), 800 (2020).
103. S. Ogawa, Y. Takagawa, and M. Kimata, "Broadband polarization-selective uncooled infrared sensors using tapered plasmonic micrograting absorbers," *Sens. Actuator A Phys.* 269, 563–568 (2018).
104. C. Fang, X. Chen, and X. Yi, "High-speed CMOS readout integrated circuit for large-scale and high-resolution microbolometer array," *Optik* 125(13), 3315–3318 (2014).
105. M. Kimata, "Uncooled infrared focal plane arrays," *IEEJ Trans. Elec. Electron. Eng.* 13, 4–12 (2018).
106. P. E. Malinowski, E. Georgitzikis, J. Maes, I. Vamvaka, F. Frazzica, J. Van Olmen, P. De Moor, P. Heremans, Z. Hens, and D. Cheyns, "Thin-film quantum dot photodiode for monolithic infrared image sensors," *Sensors* 17(12), 2867 (2017).
107. V. Pejovi, E. Georgitzikis, J. Lee, I. Lieberman, D. Cheyns, P. Heremans, and P. E. Malinowski, "Infrared colloidal quantum dot image sensors," *IEEE Trans. Electron Devices* 69(6), 2840–2850 (2022).
108. Z. M. Beiley, R. Cheung, E. F. Hanelt, E. Mandelli, J. Meitzner, J. Park, A. Pattantyus-Abraham, and E. H. Sargent, "Device design for global shutter operation in a 1.1-μm pixel image sensor and its application to near infrared sensing," *Proc. SPIE* 10098, 00981L (2017).

5 Displays

A display consists of semiconductor circuits to deliver digital information to each pixel and optical components to transform it into a visible-light pattern to be presented to an observer. Building blocks for displays include a wide range of technologies as shown in Figure 5.1. For example, a liquid crystal panel is constructed by sandwiching a liquid crystal layer with two transparent substrates. An array of thin-film transistors (TFTs) and pixel electrodes are formed on one substrate and color filters and a uniform transparent electrode are formed on the other substrate. A polarizer is attached to the external surface of each substrate. This assembly is placed on a planar light source to build a transmissive liquid crystal display (LCD). TFT arrays are used to drive organic light-emitting diodes (OLEDs) as well. For a small-area reflective LCD, a crystalline Si substrate can be used for one of the substrates. This technology is called liquid crystal on silicon (LCoS). Displays based on micro-electro-mechanical systems (MEMS) technology are also fabricated on crystalline Si substrates. For example, a laser projector can be assembled

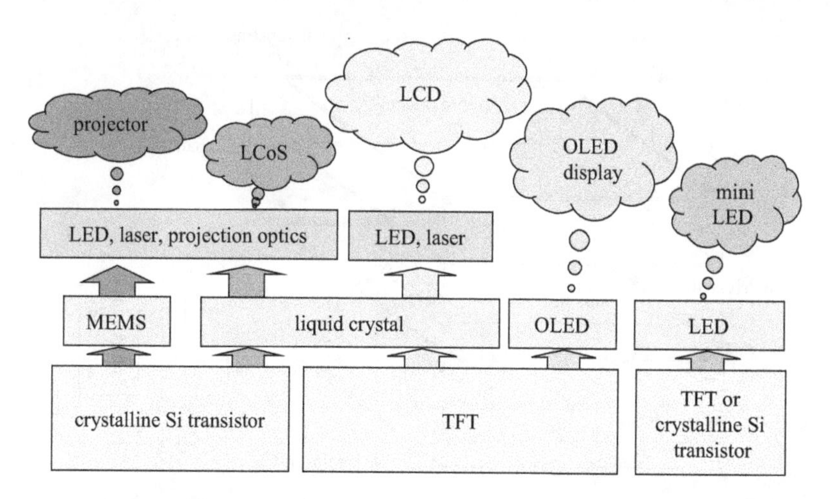

Figure 5.1 Building blocks for various displays. Note that large-area LCDs and OLED displays are driven by TFTs.

DOI: 10.1201/9781003364979-5

with MEMS devices, semiconductor laser diodes, and projection optics. Vivid color images are displayed by additive color mixing of laser light.

5.1 Liquid crystal displays

Commercialization of small-area LCDs started in the 1980s. In the 1990s, laptop computers were the primary applications of LCDs. Wide-viewing angle technologies paved the way to applications for large-area monitors and televisions. In the early 2000s, the diagonal size of LCDs reached 40 inches. The size of a mother glass on which TFTs were fabricated continued to extend. High-mobility TFTs and low-resistance metal wirings increased the panel size and pixel counts. Backlight units utilizing LEDs extended color gamut. These developments put an end to the rivalry with plasma display panels for TV applications. In the early 2020s, LCDs still dominate the market of flat panel displays, although OLED technology has started to compete since the mid-2010s.

5.1.1 Basic configuration

In a typical transmissive LCD, a liquid crystal (LC) panel is stacked on a planar light source called backlight unit as illustrated in Figure 5.2. The LC panel is a two-dimensional array of electro-optical switches. Each switch modulates the intensity of the light from the backlight unit. Because color filters (CFs) are fabricated on the top transparent substrate of the LC panel in a typical LCD, it is called CF substrate. The bottom substrate of the LC panel on which an array of pixels is fabricated is called TFT substrate. At each pixel, a TFT is connected to a transparent electrode as depicted in the circle.

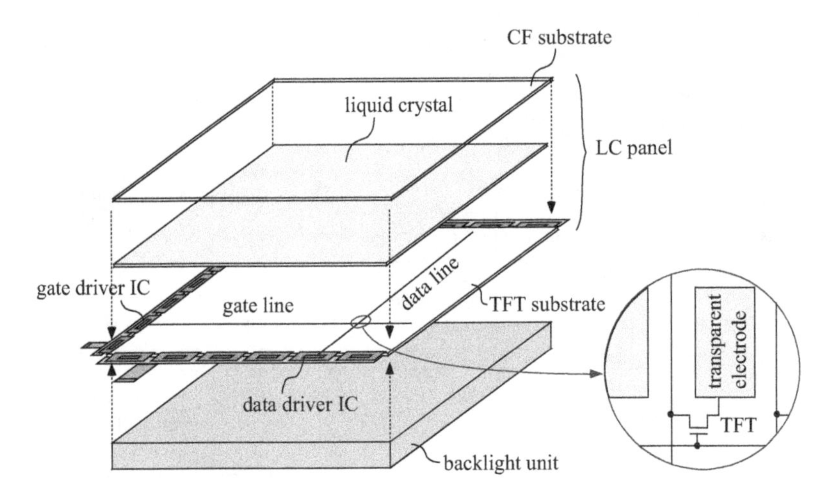

Figure 5.2 A transmissive LCD is assembled by stacking an LC panel on a backlight unit.

Two metallic lines are used to set a bias to the pixel electrode. They are called gate lines and data lines.

Depending on the LC mode of operation, the design of pixel electrodes varies. Illustrated above is an example of the TN mode and the MVA mode (see Section 2.4.2) where the uniform electrode on the CF substrate functions as a common electrode. In the case of the IPS mode, a common electrode is fabricated on the TFT substrate. Polarizer films are attached to the top and bottom surfaces of the LC panel. In addition, various optical films are used for improving optical characteristics of an LCD. For example, so-called compensation films extend viewing-angle ranges of IPS-mode LCDs [1] as well as TN-mode LCDs [2]. Reflection of ambient light at the surface of a display can be suppressed by an anti-reflection coating [3].

There is a narrow gap of about 5 μm between the two substrates. It is filled with LC materials. Originally, capillary force was utilized to inject LC materials into the narrow gap in vacuum. This LC filling process was time-consuming [4]. A large reservoir was needed for storing LC materials to fill multiple LC panels simultaneously. In 2001, these problems were solved by the technique known as one drop fill (ODF) [5]. Before assembling the two substrates, precise amount of LC material is dispensed on a TFT substrate. A sealant material, which functions as banks for the LC material later in individual LC panels, is dispensed on a counter substrate. Then, the two substrates are aligned and sealed in vacuum. Multiple LC panels are fabricated from two large substrates with a single LC filling procedure. Thus, ODF technique is fast and efficient.

A pixel of a color LCD consists of three sub-pixels as illustrated in Figure 5.3. Color filter for each color is formed on the upper substrate. A uniform transparent electrode is formed on the CFs in this example. The LC layer sandwiched by the pixel electrode and the upper electrode is equivalent to a capacitor. Orientation of the LC molecules in this region is controlled by the electric field between the two electrodes. This is not the case in the region around the TFT and the two metal lines. A light-absorbing layer

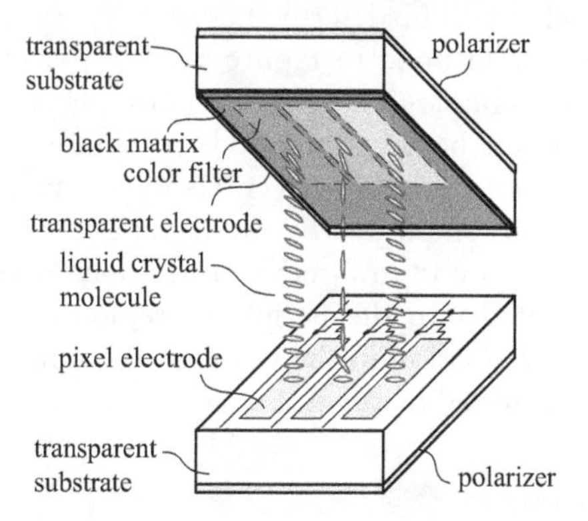

Figure 5.3 A pixel consists of three sub-pixels and each sub-pixel is addressed by two metal lines.

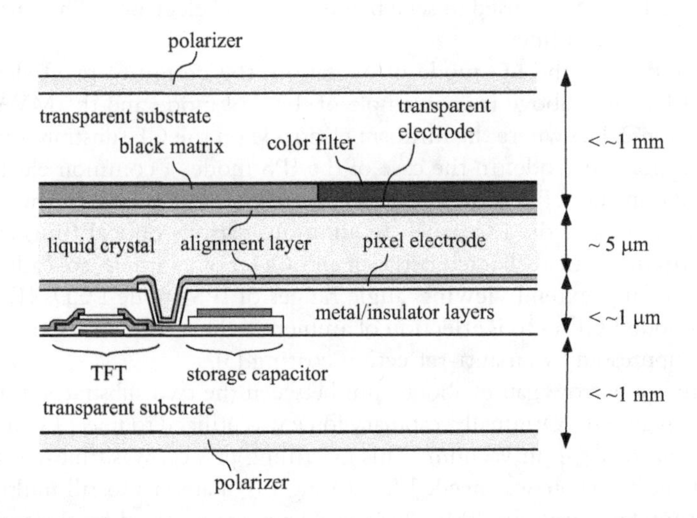

Figure 5.4 Cross section of the region around a TFT (not in scale).

called black matrix prevents the light from the backlight unit from leaking through this region.

The structure near the pixel TFT is illustrated in Figure 5.4. One terminal of the TFT is connected to the pixel electrode through the via hole made through the metal/insulator layers. The other terminals are connected to the gate line and the data line. Alignment layers are fabricated on the surfaces of the two substrates. Polarizer films are attached to the outer surfaces of these substrates. Some layers such as a retarder film and an anti-reflection coating are added (not shown). Note that the figure is not drawn to scale. Approximate thickness of each component is also shown in the figure. Also, note that an additional capacitor is formed by sandwiching insulating layers with two metal layers. This is called storage capacitor. Its objective is to increase the capacitance of a pixel as will be described next.

The TFT substrate illustrated in Figure 5.2 is an example of the case of a-Si and oxide TFT technologies. Integrated circuits (ICs) are mounted on a flexible printed circuit board. Thousands of connections must be made between the output terminals of the driver ICs and the metal lines on the TFT substrate. This task is accomplished by the technique called tape automated bonding (TAB) [6]. In case of low-temperature poly-Si (LTPS) technology, driver ICs are integrated into the peripheral region of the TFT substrate, leading to a dramatic reduction in the number of connections as well as the cost associated with external driver ICs.

5.1.2 Operation principles

An equivalent circuit for some pixels is shown in Figure 5.5a. The LC layer in each pixel is represented by a capacitor C_{LC}. The storage capacitor C_S is in

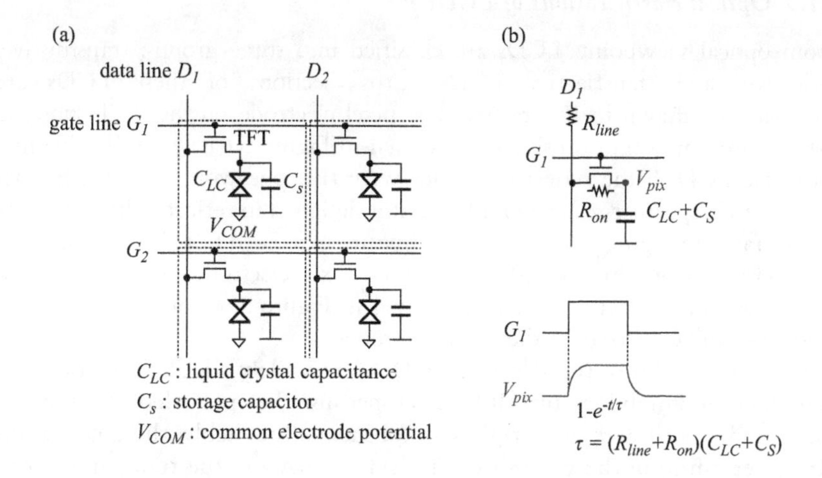

Figure 5.5 Operation of transferring image data to pixels: (a) an equivalent circuit and (b) process of charging the capacitance $C_{LC} + C_s$ through the resistor $R_{line} + R_{on}$.

parallel with C_{LC}. The bias applied to the uniform transparent electrode on the CF substrate is denoted as V_{COM}. In the case of IPS mode, the electrode for applying V_{COM} is located at each pixel on the TFT substrate.

An image is displayed by the following steps. First, data driver ICs set a certain bias pattern to the data lines, D_1, D_2,.... This corresponds to the first row of an image to be displayed. The gate driver IC turns on the TFTs in the first row. The bias pattern is stored by the pixel capacitances $C_{LC} + C_S$ in the first row. As shown in Figure 5.5b, charges are transferred through the metal line and the TFT. Denoting the resistance of the line as R_{line} and the on-time of the TFT as R_{on}, the time constant for this charging process is roughly the pixel capacitance $C_{LC} + C_s$ times the total resistance $R_{line} + R_{on}$. The pulse duration on the gate lines must be sufficiently longer than this time constant. After this charging process is completed, these TFTs are turned off by the gate driver ICs. The LC molecules reorient themselves and the first row of the image is displayed. Next, the data drivers set the bias pattern to the data lines for the second row. The gate driver ICs turn on the TFTs in the second row, and these pixels start to display the second-row image. This process is repeated for all the rows to display one image called frame. For example, if 60 frames are displayed in 1 second, each TFT is turned on every 1/60 s.

Because ions are contained in the LC layer [7], they accumulate at one region of the LC layer under a fixed bias. To mitigate this ion accumulation, the data driver ICs reverse the polarity of the bias pattern when the frame is refreshed next time. However, small amounts of ions still accumulate while the TFT is turned off. Addition of the storage capacitor reduces the voltage drop across the LC layer in this time duration.

5.1.3 Optical classifications of LCDs

From optical viewpoint, LCDs are classified into three groups: transmissive, reflective, and transflective LCDs. Cross sections of these LCDs are schematically drawn in Figure 5.6. The pixel electrode on the TFT substrate is transparent in a transmissive LCD shown in Figure 5.6a. It is reflective in a reflective LCD. If the reflective surface were flat, one would observe his/her own face via specular reflection. Hence, it is designed to reflect light diffusively as illustrated in Figure 5.6b. A periodic texture is avoided to prevent generation of Moiré patterns. In a transflective LCD, pixel electrodes are divided into transparent and reflective regions as shown in Figure 5.6c, or semitransparent materials can be utilized for the pixel electrodes.

In the early 2000s, transflective LCDs were widely adopted by notebook computers and mobile phones [8]. This is because they provided a balance for usage under both dark and bright environments. Inevitably, there is a trade-off in determining the ratio of the reflective area and the transparent area. Light utilization is compromised in both cases. In this regard, combination of a front-light unit and a reflective LCD might give a better solution [9].

5.1.4 Advanced LCDs

In 2012, a set of parameters for television, commonly known as Super HiVision (or ultra-high-definition television, 8k4k, etc.), have been standardized as Recommendation ITU-R BT. 2020 (Rec. 2020) [10]. For example, its pixel format, frame rate, and bit depth are 7,680 × 4,320, 120 Hz, and 12 bit, respectively. Advanced displays in the future should be compatible with this television standard. For mobile applications, low

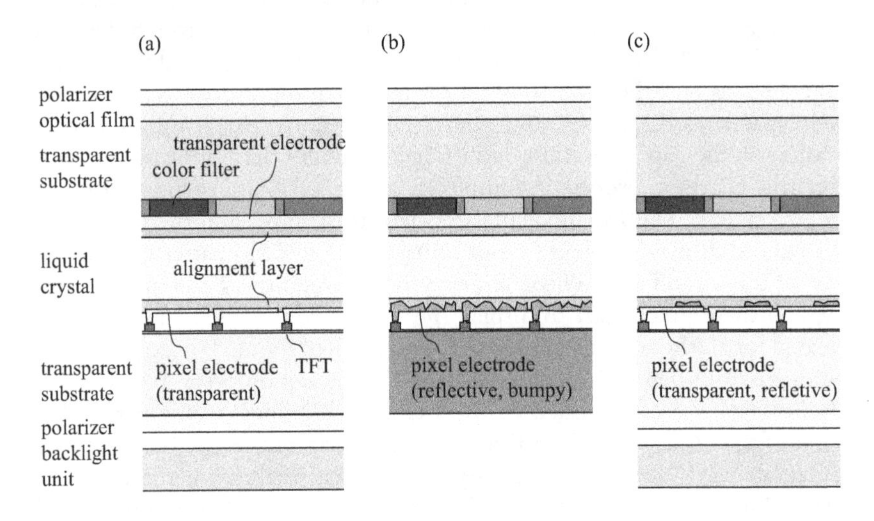

Figure 5.6 Cross sections for three types of LCDs (not in scale): (a) transmissive LCD, (b) reflective LCD, and (c) transflective LCD.

power consumption, touch input, and sunlight readability are desired features in addition to being able to display high-quality images.

5.1.4.1 Large-area LCDs

High-mobility TFTs, LC modes with a wide-viewing angle range, and other manufacturing technologies are crucial for fabricating large-area displays. For a fixed pixel format, LC capacitance at each pixel increases with the diagonal size of an LCD. The metal lines to charge and discharge pixel capacitances become longer and their resistances increase. In addition, as the frame rate is increased for displaying smooth video images, the time duration for addressing each pixel becomes shorter. Hence, the time constant for charging pixel capacitances must decrease. One way to accomplish it is to reduce the on-time resistance of TFTs. Another way is to reduce the resistance of the metal lines.

For example, an 85-inch 8k4k LCD was developed by Sharp in 2018 [11]. It adopted copper (Cu) metal lines and a-IGZO TFTs with modest mobility of 15 $cm^2/V \cdot s$. Previously, their metal lines had a double-layer structure of Cu and titanium (Ti). The Ti layer was needed to make Cu adhere to underlying materials. Diffusion of Ti atoms into the Cu layer increased the resistance of the metal lines. By eliminating Ti and somehow solving the issue of adhesion, they succeeded in reducing the resistance of the metal lines. Sharp exhibited a 120-inch LCD at a trade show in 2019.

5.1.4.2 Large-gamut LCDs

Color images are displayed by additive color mixing of the light emerging from sub-pixels. If a display utilizes three primary colors, its color gamut is represented by a triangle in CIE chromaticity diagrams. The apexes of this triangle are determined by the chromaticity coordinates of the three primary colors. They are given by integrating the product of the spectrum of the light from the sub-pixels and the CIE color matching functions (see Appendix A3).

One method to extend color gamut is to make it a polygon by adding sub-pixels for more primary colors. In 2006, Sharp reported a four-primary-color LCD [12] and commercialized it later. Conventional sub-pixels for green were replaced by those transmitting emerald-green light and yellowish-green light. Combined with a white LED backlight unit, the color gamut exceeded that of the National Television System Committee (NTSC) standard. One drawback of this approach is increased process complexity which could reduce production yield.

Another method is to narrow the spectra of the light emerging from sub-pixels. They are determined by the spectral transmittance of CFs and the spectra of light sources. One could narrow the transmittance bands of CFs. Inevitably, its luminance would decrease. Hence, narrowing the emission spectra of a backlight unit is a better option. Backlight units based on blue LEDs and inorganic phosphors for green and red sub-pixels extended the color gamut of an LCD to 96% of the NTSC standard in the CIE 1931 chromaticity diagram [13]. Backlight units utilizing quantum dots pumped by blue LEDs extended it to 115% [14].

A large color gamut is realized by monochromatic light sources because their chromaticity coordinates reside on the spectral locus, i.e., the boundary of chromaticity diagrams. In fact, the Rec. 2020 standard specifies its color gamut with monochromatic light at wavelengths of 467 nm, 532 nm, and 635 nm.

An edge-lit backlight unit utilizing laser diodes and an optical fiber was reported in 2007 [15]. An optical fiber guided the monochromatic light from a laser diode to a light-guide plate (LGP) and an output-coupler sheet in contact with the LGP extracted the light. Bright spots known as speckle noise deteriorated the uniformity of this backlight unit. This phenomenon is caused by the coherent nature of laser light. Namely, the coherent light scattered by various objects enters our eyes and constructive interference occurs on our retina, resulting in randomly distributed bright spots. By vibrating the optical fiber with an ultrasonic transmitter, different interference patterns were quickly averaged on the retina, and the speckle noise was reduced to an acceptable level [16].

Mitsubishi Electric reported a backlight unit based on red laser diodes and cyan LEDs in 2012 [17]. Its color gamut in the CIE 1976 UCS chromaticity diagram was reported to be 1.29 times larger than those of LCDs utilizing conventional white LEDs [18]. They commercialized 4k LCDs based on this technology in 2014 [19]. How the problem of speckle noise was overcome was not described. One possibility is that multiple laser diodes emitting independently average the speckle pattern. It is also possible that some moving components in a television set such as cooling fans introduce mechanical vibrations. In 2015, Mitsubishi Electric developed backlight units based on laser diodes emitting at 465 nm, 530 nm, and 639 nm. The color gamut of an LCD with this backlight unit covered 98% of the BT 2020 standard in the CIE 1976 UCS chromaticity diagram [20].

5.1.4.3 Energy-efficient LCDs

A backlight unit (BLU) consumes most of the power in a transmissive LCD. In most BLUs, multiple LEDs are placed either on a plate or the edge surface of an LGP. These are called direct-lit BLU and edge-lit BLU, respectively. In both cases, there is a one-to-one correspondence between each LED and the region on the BLU for its light emission. Hence, one can adjust the light intensity of each LED according to the images to be displayed. For example, an image of the moon in the night sky requires light output for the region corresponding to the moon. The LEDs corresponding to the dark sky regions are turned off. Thus, the power of a BLU is saved for displaying dark images. This technique is called local dimming or adaptive dimming [21].

Even for displaying a still image, driver ICs refresh it at 60 Hz for example because ions accumulated in pixels gradually cancel the charges stored in the LC capacitances. By embedding a memory in each pixel and holding the image information on the TFT substrate, one can refresh the frame much less frequently. This idea of "memory in a pixel" saves the power that otherwise would have been consumed by the driver ICs. For example, it was reported in 2002 that the frame rate for a 512-color LCD with 132 × 162 pixels was reduced to 4 Hz and that the power consumption was 1/8 of an equivalent LCD refreshed at 60 Hz [22]. In this experiment, LTPS TFTs were used to integrate complementary–metal–oxide semiconductor (CMOS) driver circuits and pixel TFTs. In 2020, combination of n-type a-IGZO TFTs and p-type LTPS TFTs was adopted for embedding a memory circuit in a pixel. Taking advantage of the low leakage current of a-IGZO TFTs, the frame rate was reduced to 1/60 Hz. The power consumption by the panel excluding a backlight unit was reduced to 0.02% of that of a conventional design [23].

5.1.4.4 In-cell touchscreen

Touchscreens are indispensable in today's mobile gadgets. Rather than stacking a separate device on a display, embedding this function inside a display has merits such as reduction of parallax error, compact form factor, and low manufacturing cost. When a finger touches an LCD, the substrates are deformed slightly. This results in a change in the LC capacitances which can be detected by fabricating sensor circuits on a TFT substrate. Multi-touch capability was demonstrated with a 2.0-inch color LCD in 2009 [24]. The sensor circuits were embedded with LTPC TFTs and the spatial resolution for touch sensing was 100 pixels per inch (ppi). Investigations on circuit designs and sensor electrode materials continued. For example, the width of bezel region for a 5.5-inch LCD with in-cell touch screen was reduced to 0.5 mm from a conventional 1.0 mm in 2016 [25].

5.2 OLED displays

5.2.1 Basic configuration

Illustrated in Figure 5.7 is a basic configuration of an OLED display and its simple two-transistor pixel circuit [26]. Data lines, gate lines, and TFT T_1 are used to charge the storage capacitor C_s in a pixel. The current in the OLED is controlled by TFT T_2 sets and accordingly its luminance. After T_1 is turned off, the storage capacitor holds the bias on the gate electrode of T_2. The light intensity of the OLED is fixed until the next frame.

There are three configurations to display color images as illustrated in Figure 5.8. In Figure 5.8a, OLEDs emitting three primary colors are fabricated on a substrate side by side. In Figure 5.8b, CFs are combined with white OLEDs. In Figure 5.8c, color conversion layers (CCLs) are used to

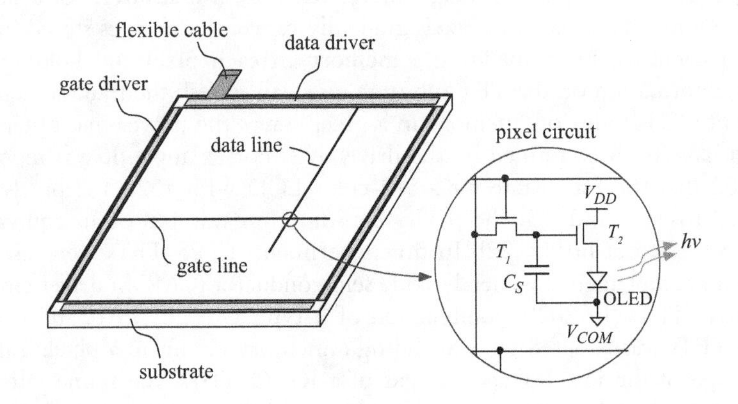

Figure 5.7 Basic configuration of an OLED display and its pixel circuit.

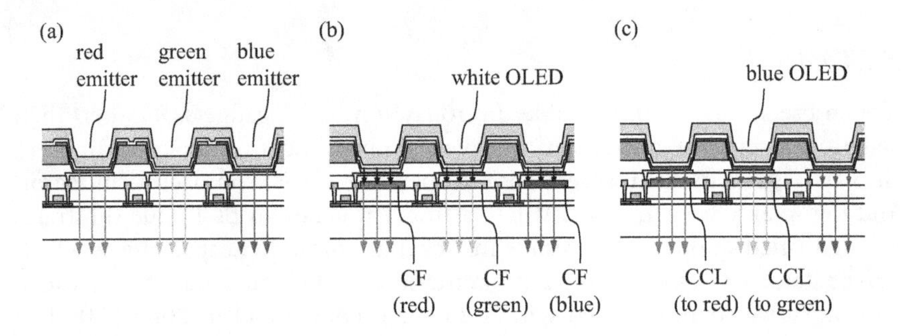

Figure 5.8 Three types of color OLED displays.

convert the blue light emitted by OLEDs to red and green light for respective sub-pixels.

Various designs have been investigated for pixel circuits to address the problems associated with TFTs such as threshold voltage shift, device-to-device variations in TFT characteristics, and leakage current. For example, a pixel circuit with four a-Si TFTs was reported in 2000 for compensating the threshold voltage shifts caused by bias stress [27]. For minimizing the area occupied by TFTs and increasing the OLED area in a pixel, high-mobility LTPS TFTs are preferred. Their drawbacks are device-to-device variations in characteristics and relatively high leakage current. The variations in mobility and threshold voltages result in non-uniform luminance. Four-TFT pixel circuits were discussed for compensating threshold voltage shifts and device-to-device variations in a-Si and LTPS TFTs [28].

Although LTPS TFTs have high mobility, it is costly to fabricate them on large substrates. In this regard, amorphous indium gallium zinc oxide (a-IGZO) TFTs offer a solution. They also have the advantage of low leakage current. Because the drain current of a MOSFET is proportional to the ratio of mobility and channel length (see Section 2.3.1), a smaller mobility can be compensated by a shorter channel. Sony reported a 9.9-inch prototype display driven by top-gate a-IGZO TFTs in 2012 [29]. A gate insulator material (SiO_2) and gate electrode materials (stacked layers of Al and Ti) were deposited on an a-IGZO layer and were patterned. A 5 nm-thick Al layer was deposited on the a-IGZO layer by using the gate electrodes as a mask. Thermal annealing at 200°C reduced the resistivity of the Al layer to a sufficiently low value to be used as source/drain regions. Short-channel TFTs (channel width of 10 μm and channel length of 4 μm) were fabricated by this self-alignment process. The field-effect mobility was 21.4 $cm^2/V \cdot s$. The pixel circuit included two TFTs and one capacitor.

In general, the number of TFTs in a pixel increases with circuit complexity, which reduces the production yield. Hence, advances in TFTs in addition to organic materials are essential for improving the characteristics of OLED displays.

5.2.2 Energy-efficient OLED displays

Emission area of a bottom-emission OLED is designed such that the light is not blocked by TFTs and metal lines. This results in reduction of the fill factor which is defined as the ratio of the emission area and the pixel area. Selection of the electrode materials allows one to extract light from either the top or bottom surface of a display. Top-emission OLED displays are preferred for a higher fill factor. Thin-film encapsulation technology plays an important role. Inorganic layers tend to have pin holes through which water vapor and oxygen penetrate. Multilayer structures of organic and inorganic materials are effective because they prolong this penetration path [30].

Low-voltage operation reduces power consumption. In 2020, bottom-gate amorphous indium zinc tin oxide (a-IZTO) TFTs were reported [31]. The high mobility of 70 $cm^2/V \cdot s$ and the threshold voltage of 1.5 V allowed low-voltage (<5 V) operation. A pixel circuit with five TFTs and two capacitors was proposed to compensate for both negative and positive threshold voltage shifts, voltage drops of power line, and OLED degradation.

Decreasing frame rate is effective for reducing the power consumed by driver circuits. Large leakage current of LTPS TFTs poses a problem for this operation. One can replace n-type LTPS TFTs with a-IGZO TFTs in a pixel circuit to take advantage of low leakage current of a-IGZO TFTs. In 2020, Sony reported a 6.39-inch OLED display with top-gate n-type a-IGZO TFTs and p-type LTPS TFTs [32]. Its pixel circuit had seven TFTs (three oxide TFTs and four LTPS TFTs) and one storage capacitor. When operated at the frame rate of 1 Hz, the total power consumed by the prototype display was reduced to 70% of the case refreshed at the conventional frame rate of 120 Hz. In 2022, Sharp reported a simplified process to connect a-IGZO TFTs to LTPS TFTs [33]. Processes of fabricating one metal layer and one organic layer can be eliminated, leading to a reduction in production cost. A commercial product of a 6.6-inch panel with 1,260 × 2,730 pixels was introduced in 2021. The frame rate could be varied from 1 Hz to 120 Hz.

5.2.3 High-contrast, wide-gamut OLED displays

Because one of the pixel electrodes in an OLED is flat and reflective, ambient light is reflected. To suppress this specular reflection, it is customary to attach a circular polarizer film on the surface of an OLED display [34]. It consists of a linear polarizer and a quarter waveplate (QWP). The angle between their optical axes is set to 45°. Unpolarized ambient becomes linearly polarized by the polarizer. After propagating the QWP, being reflected by the pixel electrode, and propagating the QWP again, its polarization plane rotates by 90°. It is absorbed by the linear polarizer (see Section 2.4.1). Therefore, ambient contrast ratio (ACR) increases with the addition of a circular polarizer. However, the linear polarizer absorbs at least 50% of the light emitted by OLEDs.

To extend color gamut of a display, narrow emission spectra are required. One can always make transmission bands of CFs narrower at the expense of luminance and related possible side effects such as increased power consumption and a shortened lifetime of organic materials. In 2006, Sony reported 100% coverage of the NTSC standard and a 1,000:1 contrast ratio in a dark environment with a configuration where a substrate with CFs was placed over three-color top-emission OLEDs. In addition, the thickness of organic layers was adjusted to minimize the reflection of ambient light for each color. It was reported that the reflection of ambient light is suppressed such that a circular polarizer could be eliminated [35]. Because this approach is based on interference of light in thin films, reflectance is expected to depend on the direction of light propagation.

5.2.4 Comparison with transmissive LCDs

Unlike an LCD, an OLED display is an emissive display. Because no backlight unit is required, it is thinner than a transmissive LCD. This is a clear advantage of OLED displays over transmissive LCDs. In addition, three aspects of their performances (response time, contrast ratio, and color gamut) are compared below.

The response of a discrete OLED is much faster than that of an LC cell in which LC molecules must change their orientations according to the change in the applied bias. In case of a display, however, it is important to remember that images are presented to an observer at a fixed interval equal to the inverse of the frame rate. Motion blur in a transmissive LCD is lessened by turning on its backlight unit once within this time interval only for a short period of time [36]. Such a backlight modulation technique has been matured by the mid-2000s [37].

Luminance of an emissive display such as an OLED display and a plasma display panel can be set to zero by not turning on its pixels. When viewed under complete darkness, its contrast ratio becomes infinity. However, a display is more likely to be viewed under ambient light. In such a case, the ACR is a more practical index for evaluating image quality.

For extending the color gamut of an OLED display, development of organic light-emitting materials continues. For example, the use of quantum dots for CCLs is investigated [38]. Only if novel materials could emit monochromatic light in future, the color gamut of an OLED display can be as wide as that of an LCD with a laser backlight unit.

5.3 LED displays

Inorganic LEDs are bright and robust. Since blue LEDs based on gallium nitride (GaN) became available, they have been widely used in general lighting and electrical appliances including backlight units for LCDs. The technology for large outdoor LED displays was established in 1999. The so-called "pick and place" technique is applied for assembling LED modules: dedicated machinery picks up LED dies from a wafer and places them on a printed circuit board. The pixel pitch is 10–50 mm for outdoor displays and 4–10 mm for indoor displays. A viewing angle of 120° is required for outdoor use, while an angle larger than 160° is desired for indoor use [39].

The size of a conventional LED die is about 300 μm × 300 μm. Smaller dies pave the way to new applications. Typically, dies smaller than 100 μm are called micro-LEDs, whereas the size of mini-LEDs is about 100 μm~200 μm. It is easier to assemble mini-LEDs than micro-LEDs.

5.3.1 Large-area displays based on micro-LEDs

In 2012, Sony demonstrated a 55-inch color LED display and commercialized a tiled version of this technology in 2016. Its pixel consists of blue, green, and red micro-LEDs and an integrated circuit for driving each LED by pulse width modulation. The size of each LED is about 20 μm and the pixel pitch is 1.26 mm. By making more than 99% of the display surface light-absorbing, they simulate that the ACR of such a display is larger than 10,000 under an illuminance of 100 lx [40]. It is speculated that massive parallel automated assembly by robots speeds up production. Modular design improves production yield and speed in general.

The demonstration by Sony in 2012 has renewed the interest in using inorganic LEDs for future displays. For example, displays in automobiles need to have high luminance and withstand high temperatures. By making the substrate of LED displays flexible, curved surfaces in cars can display images. Transparent LED displays can be used in a head-mounted display for augmented reality. Because micro-LEDs are light sources, their applications are not limited to displays. Prospects for more diversified applications of micro-LEDs have been noted [41].

5.3.2 Micro-displays

Development of a 0.5 mm × 0.5 mm micro-display was reported in 2001 [42]. An array of 12 μm-diameter p-GaN/InGaN/GaN quantum well LEDs was fabricated on a sapphire substrate. Each of the 10 × 10 LEDs had its own control pad. This was a monochrome display with passive matrix addressing.

Active matrix addressing was desired for expanding the pixel format. Although integration of driving circuits with LTPS technology on top of LED substrates was discussed in 2015 [43], most experimental studies were based on flip-chip bonding substrates with CMOS driver circuits to LED substrates. In 2011, an array of InGaN micro-LEDs on a sapphire substrate was connected to a CMOS circuit on a Si wafer by flip-chip bonding using indium bumps. The pixel format was 160 × 120, and the pixel size and pitch were 12 μm and 15 μm, respectively. A green grayscale image was displayed by actively addressing each pixel [44]. In 2013, blue and green micro-LED arrays were fabricated on sapphire substrates with GaN-based materials. Red micro-LED arrays were fabricated on a GaAs substrate and transferred to a sapphire substrate. Driver circuits on a Si wafer were connected to each of these micro-LED arrays by flip-chip bonding. The pixel format of these arrays was 60 × 60, and the pixel size and pitch were 50 μm and 70 μm, respectively [45]. A color projector was demonstrated by using these micro-LED displays.

Integrating micro-LEDs emitting three primary colors on one substrate is challenging [46]. For example, it was reported that a stacked configuration of three emitters displayed color images in 2015 [47]. In 2020, InGaN-based LEDs with four different emission spectra were fabricated side by side on one substrate rather than stacking three emitters. The LEDs had columnar structures with controlled diameters. Their peak emission wavelength varied monotonically from 478 nm to 647 nm when the diameter increased from about 100 nm to 300 nm [48].

5.3.3 Mini-LED displays

One application of mini-LEDs is a backlight unit for an LCD. Dynamic range of an LCD can be boosted by increasing the number of divisions for local dimming. This is accomplished by a direct-lit backlight unit using mini-LEDs. For example, the maximum contrast ratio of a 6.46-inch IPS-mode LCD was 3,000,000 with 288 divisions whereas that of an equivalent LCD without local dimming was 1,500 [49].

A transparent display is another application of mini-LEDs. The area not covered by TFT circuits and mini-LEDs in each pixel can be transparent. For example, a mini-LED display with 60% transmittance was reported in 2020. The size of mini-LEDs was 100 μm × 200 μm and the pixel pitch was 550 μm. The LEDs were connected to an a-IGZO TFT substrate by flip-chip bonding with solder paste. The color gamut was 114% of the NTSC standard in the CIE 1931 chromaticity diagram [50].

5.4 Sunlight-readable displays

Transmissive LCDs, OLED, and LED displays are emissive displays. They become unreadable under very bright environments. On the other hand, reflective displays such as reflective LCDs, transflective LCDs, and electrophoretic displays [51] utilize ambient light. Because the luminance of these displays is proportional to the incident light intensity, they are readable under the sun.

5.4.1 Ambient contrast ratio

Readability of a display in bright environments is quantified by the index called ambient contrast ratio (ACR). Under ambient light illuminance I_a, the light reflected by a display is added to its luminance. Suppose that the reflected light propagates uniformly over a hemisphere (solid angle of 2π sr). Denoting the reflectance of a display as R_{top}, the luminance due to the reflected ambient light is expressed as $R_{top}I_a/2\pi$. Denoting the luminance of on-time and off-time pixels as L_{on} and L_{off}, respectively, the ACR is expressed as follows [52].

$$ACR = \frac{L_{on} + R_{top}I_a/2\pi}{L_{off} + R_{top}I_a/2\pi} \tag{5.1}$$

Let us express this index for a transmissive LCD and a reflective LCD by referring to their simplified configurations in Figure 5.9.

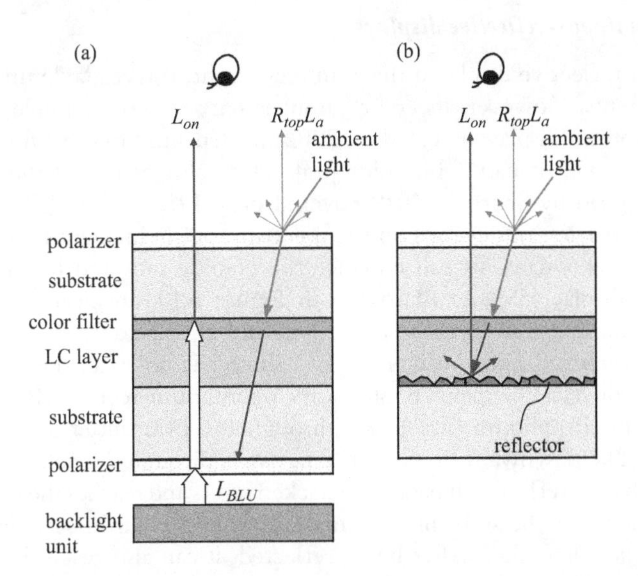

Figure 5.9 Cross sections of simplified structures for (a) a transmissive LCD and (b) a reflective LCD.

For a transmissive LCD, L_{on} is due to the light passing through its LC panel. Denoting the transmittance of the LC panel as T_{LC} and the luminance of the backlight unit as L_{BLU}, L_{on} is expressed as $T_{LC}L_{BLU}$. Let us assume that its off-state luminance is sufficiently smaller than the contribution of the light reflected at the surface, i.e., $L_{off} << R_{top}I_a/2\pi$. Then, its ACR is approximated by,

$$ACR_{T\text{-}LCD} \approx 1 + \frac{2\pi T_{LC}L_{BLU}}{R_{top}I_a} \quad (5.2)$$

As I_a increases, the second term approaches zero and the display becomes unreadable. One can increase L_{BLU} to prevent this. However, the power consumption of the backlight unit becomes prohibitively large at some point, especially for mobile applications. Nevertheless, the parameters R_{top}, T_{LC}, and L_{BLU} depend on the angle of observation.

For a reflective LCD, L_{on} is expressed as $R_{LC}I_a/2\pi$, where R_{LC} is the reflectance of the on-state pixel. Thus, its contrast ratio is given by,

$$ACR_{R\text{-}LCD} \approx 1 + \frac{R_{LC}}{R_{top}} \quad (5.3)$$

Because this is a constant larger than unity, a reflective LCD is readable even under extremely bright environments. When no auxiliary light source is used, it consumes much less power than an emissive display.

5.4.2 Luminous-reflective displays

Because a reflective display utilizes ambient light, it is readable under bright environments. However, its color gamut is narrow. For example, the color electrophoretic display reported in 2002 needed only one bit for each primary color, i.e., it had 3-bit color [53]. The color gamut of the reflective LCD reported by Sharp in 2019 covered only 19% of the NTSC standard [54]. Nevertheless, it depends on the spectrum of the incident light.

Luminescent materials can enhance the color gamut and luminance of a reflective display [55]. As illustrated in Figure 5.10, an electro-optic (EO) shutter, a luminescent layer, a color filter, and a reflector are stacked in this order for each sub-pixel. When the EO shutter is on-state, the luminescent material converts incident photons to photoluminescence (PL) photons. The upward PL photon flux passes through the EO shutter and reaches an observer. The downward PL photon flux passes through the color filter and is reflected by the reflector. It passes the stacked layers and reaches the observer. In addition, a part of the ambient light unabsorbed by the luminescent layer passes through the color filter. After being reflected, it can also reach the observer. Because both PL photons and reflected ambient light are utilized for displaying images, this configuration is called luminous-reflective display (LRD).

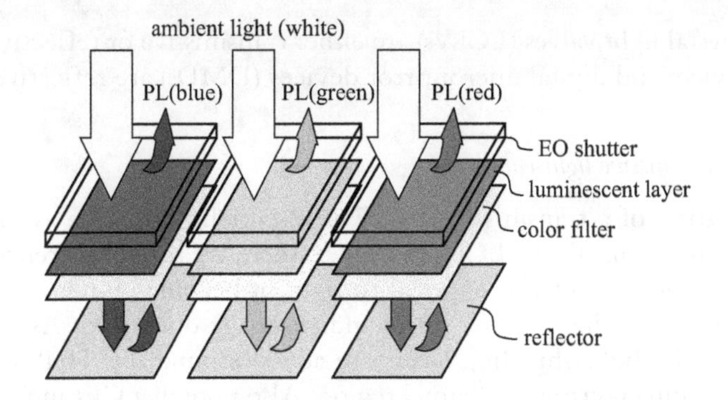

Figure 5.10 Configuration of sub-pixels in an LRD.

Luminance of an LRD is proportional to the illuminance of ambient light because the number of photoluminescent photons is proportional to the number of incident photons. Hence, its ACR is independent of illuminance. This was verified by an experiment with an LC cell stacked on a luminescent layer, a color filter, and a reflector [56].

An LRD has a larger and more stable color gamut than a reflective display because the spectra of photoluminescence are fixed irrespective of the spectra of incident photons. Furthermore, an LRD utilizes ambient light with shorter wavelengths more efficiently than a purely reflective display. For example, blue and green incident photons are converted to red PL photons at red sub-pixels in an LRD. In a reflective display, they are absorbed by the color filter at its red sub-pixel. Experiments with stacks of luminescent material/color filter/reflector revealed more than a three-fold increase in the spectral flux for red sub-pixels compared with a control device without luminescent materials [57].

5.5 Projectors

Until the mid-2000s, rear-projection televisions were popular because direct-view, large-area LCDs, and plasma display panels were relatively expensive. Today, LCDs dominate the television market. Projectors are still used for very large-screen applications such as digital cinemas and planetariums. At the other end, compact projectors are used in wearable displays for virtual/augmented reality.

To project color images on a screen, one can combine a light source, two-dimensional arrays of electro-optical switches called spatial light modulators (SLMs), and associated optical elements. Alternatively, a beam of light can be scanned on a screen with mirrors based on MEMS technology. This type of projector is almost synonymous with a laser projector.

These two types of projectors are described in the order below. In both cases, wavefront control plays an important role.

5.5.1 Projectors based on SLMs

Liquid crystal light valves (LCLVs) are either transmissive or reflective SLMs. LCoS devices and digital micromirror devices (DMDs) are reflective SLMs.

5.5.1.1 Transmissive light valves

Configuration of a transmissive LCLV is essentially the same as that of an LC panel in a transmissive LCD. However, there are some differences. If the channel region of a TFT is exposed to light, it becomes conductive. Then, the voltage across the LC layer drops and an image disappears. As illustrated in Figure 5.11, light-shielding layers are added around the TFT to prevent stray light from entering its channel region. Also note that CFs and polarizers are not present. Control of color and polarization is provided by external optical components as described below.

The TFT substrate of an early LCLV was fabricated on a quartz substrate by high-temperature poly-Si TFT technology. The diagonal size of an example reported in 1995 was 3.3 inches and it had 1,840 × 1,035 pixels [58]. The poly-Si film was formed by solid phase crystallization of an a-Si film at 600°C. The maximum process temperature reached 900°C in the following oxidation process. The pixel switch in each pixel was an n-type poly-Si TFT with a channel width/length of 3 μm/6 μm. Driver CMOS circuits were also fabricated with poly-Si TFTs. The field-effect mobility of n-type and p-type poly-Si TFTs was 160 $cm^2/V \cdot s$ and 110 $cm^2/V \cdot s$, respectively. In another example reported in 1998, the diagonal size of the LCLV was 1.8 inches with a pixel format of 1,024 × 768 [59].

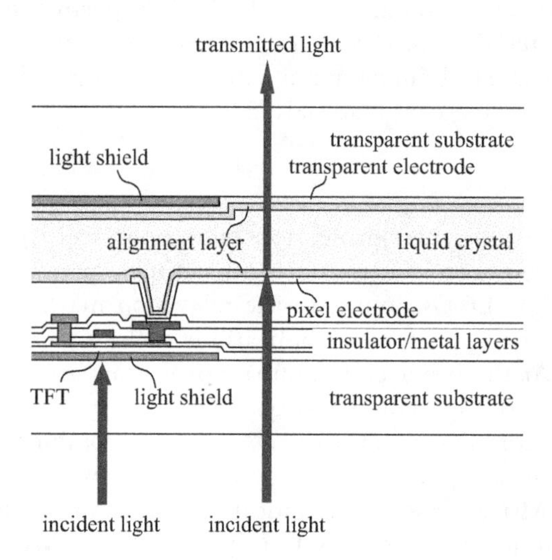

Figure 5.11 Cross section of the region around a TFT in a transmissive LCLV (not in scale). The light shield layers around the TFT prevent stray light from entering its channel region.

LTPS TFT technology allows one to use less expensive borosilicate glass substrates. A transmissive LCLV driven by LTPS TFTs was reported in 2004 [60]. The diagonal size of this LCLV was 0.9 inch with 1,024 × 768 pixels.

The ratio of the area for light transmission to that of a pixel is called fill factor or aperture ratio. Because metal lines, TFTs, and light-shielding layers are opaque, fill factor of a transmissive SLM cannot be high. For example, a projector with 1.3-inch LCLVs was commercialized in 2001. Its aperture ratio was 70% [61].

5.5.1.2 *Reflective light valves*

In 1998, a 2-inch reflective LCLV with 1,024 × 768 pixels was reported [62]. For fabricating LTPS TFTs in this early study, an Ar laser was used to crystallize a-Si films rather than an excimer laser. Organic layers were used for planarization as well as for absorbing stray light.

The bottom substrate of a reflective light valve can be a crystalline Si substrate. A DMD is an array of oscillating mirrors fabricated on Si substrates with MEMS technology (see Section 2.4.3). Texas Instruments named a projection system based on DMDs as a Digital Light Projector (DLP) [63]. Alternatively, one can fabricate a reflective light valve by modulating light with LC materials. This technology is called LCoS [64]. By 2013, LCoS devices with 8k × 4k resolution and pixel size of 4.8 μm were demonstrated. The gap between neighboring pixel mirrors was 0.2 μm, resulting in a fill factor of 93% [65]. Compared to transmissive LCLVs, DMDs, and LCoS devices have a higher fill factor because their pixel circuits can be fabricated under reflectors. Both DLP and LCoS technologies are dominant in large-screen projector applications.

Driver circuits for DMDs and LCoS devices are fabricated on Si substrates by standard CMOS processes. It is important to shield MOSFETs from stray light from an intense light source. For example, the cross-sectional view of the pixel region of an LCoS device is illustrated in Figure 5.12. Multiple light-shielding layers are placed beneath the gap between pixel electrodes.

5.5.1.3 *Projector configurations*

Various types of projectors are configured depending on how many SLMs are used, whether the SLMs are transmissive or reflective, how the polarization state is controlled if required, and so on. An example using three LCLVs is illustrated in Figure 5.13. A white light source emits intense unpolarized light. The mirror around it makes it a parallel beam. A pair of microlens arrays homogenizes its intensity distribution. The next optical component consists of an array of a polarization beam splitter (PBS) and a half-waveplate (HWP). A PBS is a pair of prisms sandwiching periodic thin films with alternating refractive indices. It is designed to pass p-polarized light and reflect s-polarized light [66]. Combination of a PBS and a HWP converts unpolarized

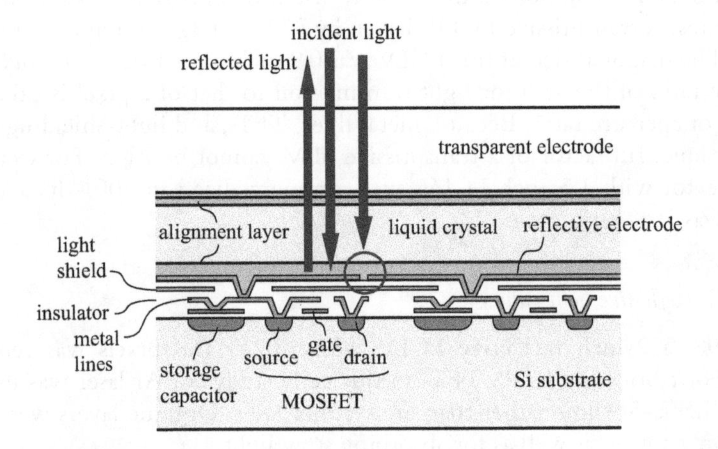

Figure 5.12 Cross section of the pixel region in an LCoS device (not in scale). Multiple light-shielding layers below the gap between pixel electrodes block stray light from an intense light source (see the circled region at the center of this drawing).

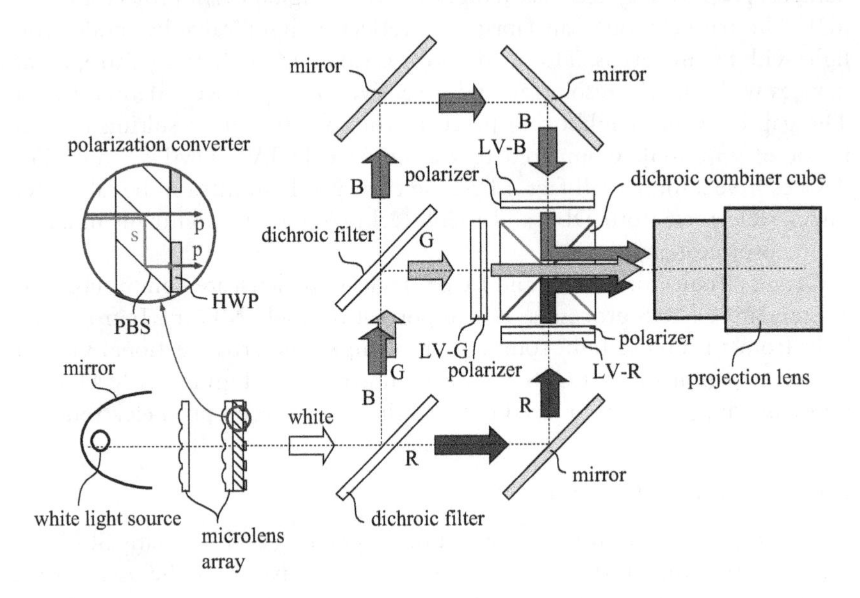

Figure 5.13 Optical configuration for a color projector utilizing three transmissive LCLVs. (PBS; polarization beam splitter, HWP; half-waveplate, LV-R, LV-G, LV-B; light valve for red, green, and blue light).

light to p-polarized light. Namely, as shown in the red circle, p-polarized light passes the PBS while s-polarized light is reflected. It is reflected by a neighboring PBS and the reflected light enters the HWP. Its polarization plane is rotated by 90° and p-polarized light emerges from the HWP. Thus,

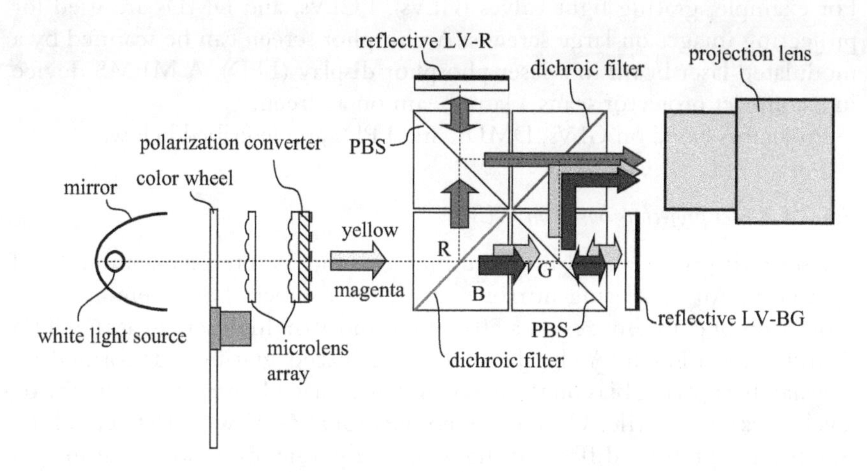

Figure 5.14 A color projector utilizing two reflective LCLVs and a color wheel.

the white light entering the first dichroic filter is linearly polarized. For this reason, this component is called polarization converter.

The first dichroic filter transmits the red light and reflects green and blue light. The second dichroic mirror separates the green and blue light. Hence, the red, green, and blue lights enter the respective LCLVs denoted as LV-R, LV-G, and LV-B. Each LCLV and the polarizer placed at its exit surface modulate the intensity of the incident light. The dichroic combiner cube consists of two types of dichroic filters. It reflects red and blue light and transmits green light. The combined light is projected on a screen by the projection lens.

An example of the configuration based on two reflective LCLVs is shown in Figure 5.14. A rotating color wheel transmits yellow and magenta light alternately. Red light always exists in the transmitted light while green and blue light is interchanged with time. They are all linearly polarized by the polarization converter. The first dichroic filter reflects the red light. It transmits the blue and green light in turn. The reflective LV-R and the PBS nearby modulate the intensity of the red light. Green and blue image data are sent to the reflective LV-BG in synchronous with the rotating color wheel such that blue and green images are sent to the second dichroic filter alternately. Red and green LCLVs generate a yellow image, while red and blue LCLVs generate a magenta image. They are sent to the projection lens in turn.

A light source and SLMs can be replaced by emissive displays such as OLED displays [67] and micro-LED displays [68]. This approach simplifies the associated optics and improves light utilization efficiency.

5.5.2 Laser projectors

Additive mixing of laser light allows one to display vivid color images. Various wavefront-control devices manipulate its propagation (see Section 2.4).

For example, grating light valves (GLVs), LCLVs, and DMDs are used for projecting images on large screens. A phosphor screen can be scanned by a modulated laser beam in a laser phosphor display (LPD). A MEMS device in a compact projector scans a laser beam on a screen.

Projectors based on GLVs, DMDs, and LPDs are described below.

5.5.2.1 Laser projectors based on GLVs

A laser beam is expanded in one dimension and its intensity is modulated by a GLV. An oscillating mirror scans it on a screen. Sony demonstrated projection of color images on a 50 m wide and 10 m high screen at the 2005 World Exposition in Aichi, Japan [69]. A blazed grating was formed by alternately applying bias on the array of stripe-shaped reflectors with a two-level thickness profile. Unlike the conventional GLV with flat reflectors, the blazed grating diffracted light in only one direction, leading to simplified optics. Using three lasers emitting at 445 nm, 532 nm, and 642 nm, a color gamut of 140% of the NTSC standard was realized in 2005.

A downsized version of this system with an improved contrast ratio and frame rate was reported in 2009. Several techniques were adopted to reduce the speckle contrast to 7% [70]. Speckle contrast is defined as the ratio of the standard deviation and the average intensity of an image. Because interference of laser light occurs on the retina in our eyes, measurement must mimic real-life observation conditions.

5.5.2.2 Laser projectors based on LCLVs and DMDs

Matsushita Electric Industrial Co., Ltd. reported a rear projector prototype utilizing three 0.7-inch transmissive LCLVs in 2006. They used red and blue laser diodes and a green diode-pumped solid-state laser. A rod integrator was used to homogenize the intensity distribution of the laser light. To suppress speckle noise, a rotating lens array was inserted between the lasers and the rod integrator. The emission wavelengths were 445 nm, 532 nm, and 640 nm. The color gamut was 137% of the NTSC standard in the CIE 1931 chromaticity diagram [71].

In 2008, Mitsubishi Electric commercialized a 65-inch rear projector based on DMD technology. A compact projection engine was developed based on an aspherical mirror [72], resulting in a housing depth of 255 mm. Optical fibers were used to relay the light from three lasers. Wavelengths of these lasers were 447 nm, 532 nm, and 640 nm [73].

Adding lasers emitting at different wavelengths extends the color gamut further. Such an experiment was reported in 2021. The original light source in a digital movie projector was replaced by six lasers emitting at 445 nm, 465 nm, 520 nm, 550 nm, 638 nm, and 660 nm. The light was projected on a white wall in a dark room and a spectral radiance colorimeter measured chromaticity coordinates. The measured color gamut was 178.4% of the NTSC standard in the CIE 1931 chromaticity diagram [74].

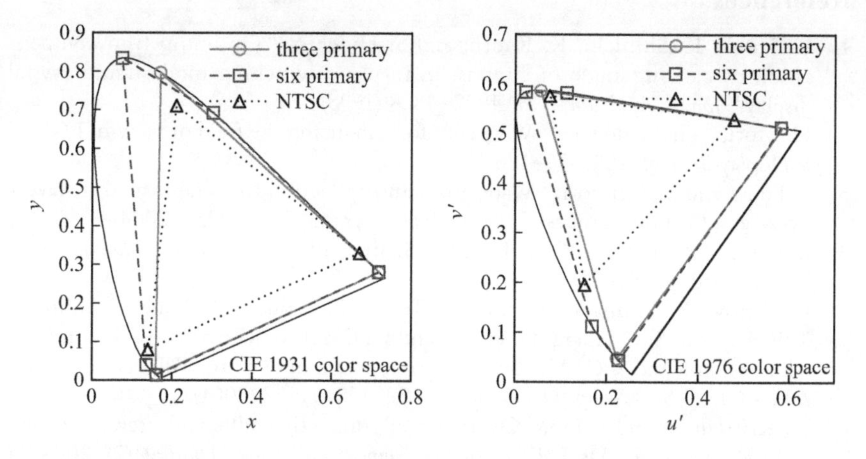

Figure 5.15 Color gamut calculated for three and six monochromatic primary colors.

Assuming monochromatic spectra at these six wavelengths, the color gamut calculated in two CIE color spaces is shown by the broken hexagon in each plot in Figure 5.15. Also shown is the color gamut calculated for the 2008 Mitsubishi product (solid triangle) and that of the NTSC standard (dotted triangle). Color gamut calculated for the three primary colors (Mitsubishi 2008 product) is 137% and 183% of the NTSC standard in the CIE 1931 and 1976 color space, respectively. Because our eyes are less sensitive to minute differences in green colors, it is more appropriate to compare color gamut in the CIE 1976 color space. The lasers reproduce red and blue colors notably well.

5.5.2.3 Laser phosphor displays

In a cathode ray tube (CRT), a phosphor screen is scanned by an intensity-modulated electron beam in vacuum to display an image. One can replace the electron beam with a violet laser beam. No vacuum is required in this case. An array of phosphor stripes is formed on a screen for emitting three primary colors. To control the luminance of each pixel, intensity of the laser beam is modulated according to video data. Two mirrors are used to scan the beam in horizontal and vertical directions. For precise alignment of the beam and the screen, retroreflectors are placed on the screen and an infrared laser beam scans the phosphor screen. By detecting the infrared light reflected by the screen, the position of the scanning engine is adjusted with a servo system. Modular design allows one to construct a large display to be used in indoor environments. This technology called laser phosphor display was reported in 2012 [75]. A rollable version of the phosphor screen and an improved beam alignment system were reported in 2018 [76].

References

1. Y. Saitoh, K. Shinichi, K. Kaoru, and S. Hidehisa, "Optimum film compensation of viewing angle of contrast in in-plane-switching-mode liquid crystal display," *Jpn. J. Appl. Phys.* 37, 4822–4828 (1998).
2. H. Mori, "The wide view (WV) film for enhancing the field of view of LCDs," *J. Display Technol.* 1(2), 179–186 (2005).
3. S. Fujita and S. Shiratori, "Waterproof anti reflection films fabricated by layer-by-layer adsorption process," *Jpn. J. Appl. Phys.* 43, 2346–2351 (2004).
4. J. Souk, S. Morozumi, F-C. Luo, and I. Bita, *Flat Panel Display Manufacturing*, Wiley, 2018, pp. 59–71.
5. H. Kamiya, K. Tajima, K. Toriumi, K. Terada, H. Inoue, T. Yokoue, N. Shimizu, T. Kobayashi, S. Odahara, G. Hougham, C. Cai, J. H. Glownia, R. J. von Gutfeld, R. John, and S-C. Alan Lien, "Development of one drop fill technology for AM-LCDs," *SID Symp. Dig. Tech. Pap.* 32, 1354–1357 (2001).
6. H. Kristiansen and J. Liu, "Overview of conductive adhesive interconnection technologies for LCDs," *IEEE Trans. Compon. Packaging Manuf.* 21(2), 208–214 (1998).
7. S. Naemura, "Liquid-crystal-material technologies for advanced display applications," *J. Soc. Inf. Display* 8(1), 5–9 (2000).
8. X. Zhu, Z. Ge, T. X. Wu, and S-T. Wu, "Transflective liquid crystal displays," *J. Display Technol.* 1(1), 15–29 (2005).
9. A. Tanaka, "Technical trends of front-lighting systems in reflective color LCD modules," *SID Symp. Dig. Tech. Pap.* 33, 1240–1243 (2002).
10. M. Sugawara, M. Emoto, K. Masaoka, Y. Nishida, and Y. Shishikui, "SUPER Hi-VISION for the next generation television," *ITE Trans. Media Technol. Appl.* 1(1), 27–33 (2013).
11. Y. Hara, T. Kikuchi, H. Kitagawa, J. Morinaga, H. Ohgami, H. Imai, T. Daitoh, and T. Matsuo, "IGZO-TFT technology for large-screen 8K display," *J. Soc. Inf. Display* 26(3), 169–177 (2018).
12. E. Chino, K. Tajiri, H. Kawakami, H. Ohira, K. Kamijo, H. Kaneko, S. Kato, Y. Ozawa, T. Kurumisawa, K. Inoue, K. Endo, H. Moriya, T. Aragaki, and K. Murai, "Development of wide-color-gamut mobile displays with four-primary-color LCDs," *SID Symp. Dig. Tech. Pap.* 37, 1221–1224 (2006).
13. L. Wang, X. Wang, T. Kohsei, K-i. Yoshimura, M. Izumi, N. Hirosaki, and R-J. Xie, "Highly efficient narrow-band green and red phosphors enabling wider color-gamut LED backlight for more brilliant displays," *Opt. Express* 23, 28707–28717 (2015).
14. Z. Luo, Y. Chen, and S-T. Wu, "Wide color gamut LCD with a quantum dot backlight," *Opt. Express* 21, 26269–26284 (2013).
15. Y. Inaba, Y. Nagai, and I. Fujieda, "Edge-lit backlight utilizing a laser diode and an optical fiber," *Proc. 14th Int. Display Workshops*, 705–708 (2007).
16. I. Fujieda, T. Kosugi, and Y. Inaba, "Speckle noise evaluation and reduction of an edge-lit backlight system utilizing laser diodes and an optical fiber," *J. Display Technol.* 5, 414–417 (2009).
17. E. Niikura, R. Murase, S. Kagawa, N. Nakano, A. Nagase, H. Sakamoto, T. Sasagawa, K. Minami, H. Sugiura, K. Shimizu, and M. Hanai, "Development of laser backlighting LCD television," *Proc. 19th Int. Display Workshops*, 287–288 (2012).

18. K. Minami, "A wide color gamut display using laser light sources," *SID Symp. Dig. Tech. Pap.* 45, 839–841 (2014).
19. N. Okimoto, S. Maeda, E. Niikura, T. Sawanaka, and H. Kida, "Development of a laser optical system for a 4K laser backlit LCD TV," *SID Symp. Dig. Tech. Pap.* 46, 1067–1069 (2015).
20. E. Niikura, N. Okimoto, S. Maeda, H. Yasui, A. Heishi, S. Yamanaka, T. Sasagawa, Y. Nishida, and Y. Kusakabe, "Development of RGB laser backlit liquid crystal display," *Proc. of 22nd International Display Workshops*, 2015, pp. 1096–1099,
21. T. Shiga, S. Kuwahara, N. Takeo, and S. Mikoshiba, "Adaptive dimming technique with optically isolated lamp groups," *SID Symp. Dig. Tech. Pap.* 36, 992–995 (2005).
22. H. Tokioka, M. Agari, M. Inoue, T. Yamamoto, H. Murai, and H. Nagata, "Low-power-consumption TFT-LCD with dynamic memory embedded in the pixels," *J. Soc. Inf. Display* 10(2), 123–126 (2002).
23. J. Kim, W-R. Lee, H-J. Chung, and S-W. Lee, "A low-power memory-in-pixel circuit for liquid crystal displays comprising low-temperature poly-silicon and oxide thin-film transistors," *Electronics* 9, 1958 (2020).
24. E. Kanda, T. Eguchi, Y. Hiyoshi, T. Chino, Y. Tsuchiya, T. Iwashita, T. Ozawa, T. Miyazawa, and T. Matsumoto, "Active-matrix sensor in AMLCD detecting liquid-crystal capacitance with LTPS-TFT technology," *J. Soc. Inf. Display* 17(2), 79–85 (2009).
25. Y. Teranishi, K. Noguchi, H. Mizuhashi, K. Ishizaki, H. Kurasawa, and Y. Nakajima, "New in-cell capacitive touch panel technology with low resistance material sensor and new driving method for narrow dead band display," *SID Symp. Dig. Tech. Pap.* 47, 502–505 (2016).
26. M. Stewart, R. S. Howell, L. Pires, M. K. Hatalis, W. Howard, and O. Prache, "Polysilicon VGA active matrix OLED display-technology and performance," *Proc. IEEE Int. Electron Devices Meet. (IEDM)*, 871–874 (1998).
27. Y. He, R. Hattori, and J. Kanicki, "Current-source a-Si:H thin-film transistor circuit for active-matrix organic light-emitting displays," *IEEE Electron Device Lett.* 21(12), 590–592 (2000).
28. A. Nathan, G. Reza Chaji, and S. J. Ashtiani, "Driving schemes for a-Si and LTPS AMOLED displays," *J. Display Technol.* 1, 267–277 (2005).
29. N. Morosawa, Y. Ohshima, M. Morooka, T. Arai, and T. Sasaoka, "A novel self-aligned top-gate oxide TFT for AM-OLED displays," *SID Symp. Dig. Tech. Pap.* 42, 479–482 (2012).
30. D. Yu, Y-Q. Yang, Z. Chen, Y. Tao, and Y-F. Liu, "Recent progress on thin-film encapsulation technologies for organic electronic devices," *Opt. Commun.* 362, 43–49 (2016)
31. C-L. Fan, H-Y. Tsao, C-Y. Chen, P-C. Chou, and W-Y. Lin, "New low-voltage driving compensating pixel circuit based on high-mobility amorphous indium-zinc-tin-oxide thin-film transistors for high-resolution portable active-matrix OLED displays," *Coatings* 10(10), 1004 (2020).
32. R. Yonebayashi, K. Tanaka, K. Okada, K. Yamamoto, K. Yamamoto, S. Uchida, T. Aoki, Y. Takeda, H. Furukawa, K. Ito, H. Katoh, and W. Nakamura, "High refresh rate and low power consumption AMOLED panel using top-gate n-oxide and p-LTPS TFTs," *J. Soc. Inf. Display* 28(4), 350–359 (2020).

33. M. Honjo, Y. Takeda, M. Aman, S. Kobayashi, K. Kitoh, K. Ito, K. Tanaka, H. Matsukizono, and W. Nakamura, "Advanced hybrid process with back contact IGZO-TFT," *J. Soc. Inf. Display* 30(5), 471–481 (2022).
34. R. Singh, K. N. Narayanan Unni, A. Solanki, Deepak, "Improving the contrast ratio of OLED displays: An analysis of various techniques," *Opt. Mater.* 34(4), 716–723 (2012).
35. T. Ishibashi, J. Yamada, T. Hirano, Y. Iwase, Y. Sato, R. Nakagawa, M. Sekiya, T. Sasaoka, and T. Urabe, "Active matrix organic light Emitting diode display based on "Super Top Emission" technology," *Jpn. J. Appl. Phys.* 45(5B), 4392–4395 (2006).
36. J. Hirakata, A. Shingai, Y. Tanaka, K. Ono, and T. Furuhashi, "Super-TFT-LCD for moving picture images with the blink backlight system," *SID Symp. Dig. Tech. Pap.* 32, 990–993 (2001).
37. C. T. Liu, "Revolution of the TFT LCD Technology," *J. Disp. Technol.* 3(4), 342–350 (2007).
38. X. Dai, Y. Deng, X. Peng, and Y. Jin, "Quantum-dot light-emitting diodes for large-area displays: Towards the dawn of commercialization," *Adv. Mater.* 29, 1607022 (2017).
39. F. Nguyen, "Challenges in the design of a RGB LED display for indoor applications," *Synth. Met.* 122(1), 215–219 (2001).
40. G. Biwa, A. Aoyagi, M. Doi, K. Tomoda, A. Yasuda, and H. Kadota, "Technologies for the Crystal LED display system," *J. Soc. Inf. Display* 29(6), 435–445 (2021).
41. J. Y. Lin and H. X. Jiang, "Development of microLED," *Appl. Phys. Lett.* 116, 100502 (2020).
42. H. X. Jiang, S. X. Jin, J. Li, J. Shakya, and J. Y. Lin, "III-nitride blue microdisplays," *Appl. Phys. Lett.* 78, 1303 (2001).
43. B. R. Tull, Z. Basaran, D. Gidony, A. B. Limanov, J. S. Im, I. Kymissis, and V. W. Lee, "High brightness, emissive microdisplay by integration of III-V LEDs with thin film silicon transistors," *SID Symp. Dig. Tech. Pap.* 46, 375–377 (2015).
44. J. Day, J. Li, D. Y. C. Lie, C. Bradford, J. Y. Lin, and H. X. Jiang, "III-Nitride full-scale high-resolution microdisplays," *Appl. Phys. Lett.* 99, 031116 (2011).
45. Z. J. Liu, W. C. Chong, K. M. Wong, and K. M. Lau, "360 ppi flip-chip mounted active matrix addressable light emitting diode on silicon (LEDoS) micro-displays," *J. Disp. Technol.* 9(8), 678–682 (2013).
46. F. Templier, "GaN-based emissive microdisplays: A very promising technology for compact, ultra-high brightness display systems," *J. Soc. Inf. Display* 24(11), 669–675 (2016).
47. H. S. El-Ghoroury, C-L. Chuang, and Z. Y. Alpaslan, "Quantum photonic imager (QPI): A novel display technology that enables more than 3D applications," *SID Symp. Dig. Tech. Pap.* 46, 371–374 (2015).
48. K. Kishino, N. Sakakibara, K. Narita, and T. Oto, "Two-dimensional multicolor (RGBY) integrated nanocolumn micro-LEDs as a fundamental technology of micro-LED display," *Appl. Phys. Express* 13, 014003 (2020).
49. Z. Deng, B. Zheng, J. Zheng, L. Wu, W. Yang, Z. Lin, P. Shen, and J. Li, "High dynamic range incell LCD with excellent performance," *SID Symp. Dig. Tech. Pap.* 49, 996–998 (2018).
50. Y. Sun, J. Fan, M. Liu, L. Zhang, B. Jiang, M. Zhang, and X. Zhang, "Highly transparent, ultra-thin flexible, full-color mini-LED display with

indium–gallium–zinc oxide thin-film transistor substrate," *J. Soc. Inf. Display* 28(12), 926–935 (2020).
51. B. Comiskey, J. D. Albert, H. Yoshizawa, and J. Jacobson, "An electrophoretic ink for all-printed reflective electronic displays," *Nature* 394, 253–255 (1998).
52. J. A. Dobrowolski, Brian T. Sullivan, and R. C. Bajcar, "Optical interference, contrast-enhanced electroluminescent device," *Appl. Opt.* 31, 5988–5996 (1992).
53. G. Duthaler, J. Au, M. Davis, H. Gates, B. Hone, A. Knaian, E. Pratt, K. Suzuki, S. Yoshida, and M. Ueda, "Active-matrix color displays using electrophoretic ink and color filters," *SID Symp. Dig. Tech. Pap.* 33, 1374–1377 (2002).
54. H. Hakoi, M. Ni, J. Hashimoto, T. Sato, S. Shimada, K. Minoura, A. Itoh, K. Tanaka, H. Matsukizono, and M. Otsubo, "High-performance and low-power full color reflective LCD for new applications," *SID Symp. Dig. Tech. Pap.* 50(1), 279–282 (2019).
55. G. Gibson, X. Sheng, D. Henze, S-T. Lam, P. Beck, Y. Jeon, Z-L. Zhou, B. Benson, Q. Liu, G. Combs, T. Koch, and K. Biggs, "Fast full-color reflective display via photoluminescent enhancement," *J. Soc. Inf. Display* 20(10), 552–558 (2012).
56. Y. Yamada, Y. Tsutsumi, and I. Fujieda, "Ambient contrast ratio of a liquid crystal cell stacked on a luminescent layer," *Proc. of 27th International Display Workshops*, 27, 2020, pp. 76–79.
57. I. Fujieda, Y. Tsutsumi, and S. Matsuda, "Spectral study on utilizing ambient light with luminescent materials for display applications," *Opt. Express* 29, 6691–6702 (2021).
58. H. Sato, H. Nakamura, Y. Masuda, T. Nakazono, M. Kobayashi, K. Mori, and N. Harada, "A 1.9 M-pixel poly-Si TFT-LCD for HD and computer-data projectors," *IEEE Trans. Consum. Electron.* 41(4), 1181–1188 (1995).
59. K. Uchino, T. Kashima, F. Abe, M. Hirano, M. Satoh, and T. Maekawa, "Ultra-high brightness 1.8-in. XGA poly-Si TFT LCD," *SID Symp. Dig. Tech. Pap.* 29, 1067–1068 (1998).
60. Y. Tomihari, K. Yoshinaga, H. Sekine, M. Sugimoto, T. Okumura, N. Seko, N. Takada, H. Okumura, K. Shiota, K. Hirata, N. Matsunaga, K. Sera, F. Okumura, "A low temperature poly-Si TFT liquid crystal light valve (LCLV) with a novel light-shielding structure for high performance projection displays," *SID Symp. Dig. Tech. Pap.* 35, 972–975 (2004).
61. N. Okamoto, "Developments in p-Si TFT LCD projectors," *SID Symp. Dig. Tech. Pap.* 32, 1176–1179 (2001).
62. M. Kunigita, N. Kato, K. Masumo, and M. Yuki, "A low-temperature poly-Si TFT reflective XGA array for LCPC light valve," *SID Symp. Dig. Tech. Pap.* 29, 463–466 (1998).
63. L. J. Hornbeck, "Digital light processing for high-brightness high-resolution applications," *Proc. SPIE* 3013, 1–14 (1997).
64. R. L. Melcher, M. Ohhata, and K. Enami, "High information content display based on reflective LC on Silicon light valves," *SID Symp. Dig. Tech. Pap.* 29, 25–28 (1998).
65. W. P. Bleha Jr. and L. A. Lei, "Advances in liquid crystal on silicon (LCOS) spatial light modulator technology," *Proc. SPIE* 8736, 87360A (2013).
66. L. Li and J. A. Dobrowolski, "Visible broadband, wide-angle, thin-film multilayer polarizing beam splitter," *Appl. Opt.* 35, 2221–2225 (1996).
67. C. Großmann, S. Riehemann, G. Notni, and A. Tünnermann, "OLED-based pico-projection system," *J. Soc. Inf. Display* 18(10), 821–826 (2010).

68. Z. J. Liu, W. C. Chong, K. M. Wong, K. H. Tam, and K. M. Lau, "A novel BLU-free full-color LED projector using LED on silicon micro-displays," *IEEE Photonics Technol. Lett.* 25(23), 2267–2270 (2013).
69. H. Tamada, "Blazed GxLP™ light modulators for laser projectors," *J. Soc. Inf. Display* 15(10), 817–823 (2007).
70. H. Kikuchi, S. Hashimoto, S. Tajiri, T. Hayashi, Y. Sugawara, M. Oka, Y. Akiyama, A. Nakamura, and N. Eguchi, "High-pixel-rate grating-light-valve laser projector," *J. Soc. Inf. Display* 17(3), 263–269 (2009).
71. T. Mizushima, H. Furuya, K. Mizuuchi, T. Yokoyama, A. Morikawa, K. Kasazumi, T. Itoh, A. Kurozuka, K. Yamamoto, S. Kadowaki, and S. Marukawa, "Laser projection display with low electric consumption and wide color gamut by using efficient green SHG laser and new illumination optics," *SID Symp. Dig. Tech. Pap.* 37, 1681–1684 (2006).
72. M. Kuwata, T. Sasagawa, K. Kojima, J. Aizawa, A. Miyata, S. Shikama, and H. Sugiura, "Projection optical system for a compact rear projector," *J. Soc. Inf. Display* 14(2), 199–206 (2006).
73. H. Sugiura, T. Sasagawa, A. Michimori, E. Toide, T. Yanagisawa, S. Yamamoto, Y. Hirano, M. Usui, S. Teramatsu, and J. Someya, "65-inch, super slim, laser TV with newly developed laser light sources," *SID Symp. Dig. Tech. Pap.* 39, 854–857 (2008).
74. L. Zhu, G. Wang, Y. Yang, B. Yao, C. Gu, and L. Xu, "Six-primary-laser projection display system: Demonstration and stereo color gamut measurement," *Opt. Express* 29, 43885–43898 (2021).
75. R. A. Hajjar, "Introducing scalable, freeform, immersive, high-definition laser phosphor displays," *SID Symp. Dig. Tech. Pap.* 43, 985–988 (2012).
76. R. A. Hajjar, "Seamless scalable large format display," *SID Symp. Dig. Tech. Pap.* 49, 737–739 (2018).

6 Miscellaneous applications

Insight into future lifestyles propels innovation. Armed with laws of physics, one can combine electronic and photonic devices in a unique manner for novel imaging and display applications. Values of specific inventions might fade as competing technologies mature and lifestyles change with time. However, configurations and operation principles of devices developed for one particular purpose might become building blocks for other objectives. Three examples in this chapter show how new concepts were perceived by the demand of society and the technologies available at the time.

The first application is a compact document scanner developed by NEC in the 1990s. Research on low-temperature poly-Si (LTPS) TFT technology had been active in developing LCDs. For justifying investment in production facilities, other possible applications were explored. A blue LED had just become commercially available. It was the missing piece for realizing a compact white light source. By developing dedicated optics, every component of a hand-held scanner was enclosed in a 10 mm-diameter cylindrical container.

The second application is an anti-spoofing technique for fingerprint imaging. Electronic commerce started to demand online authentication in the late 1990s. A prototype mobile phone with an embedded fingerprint sensor was reported in 1998. Fujitsu commercialized such a phone in 2003 [1]. With flourishing electronic commerce, the threat of spoofing increased. Clearly, anti-spoofing measures, preferably without extra hardware, were desired for secure transactions.

The third application is an energy-harvesting display. Threat of climate change calls for renewable energy sources more than ever. Two types of configurations are described for incorporating photovoltaic technologies into displays.

6.1 Compact color scanners

The structure of a pen-shaped scanner [2] is illustrated in Figure 6.1. A light source consists of an array of blue, green, and red LEDs and a cylindrical lens. A linear sensor fabricated on a glass substrate by LTPS TFT technology has an array of a-Si photodiodes and complementary metal–oxide–semiconductor

DOI: 10.1201/9781003364979-6

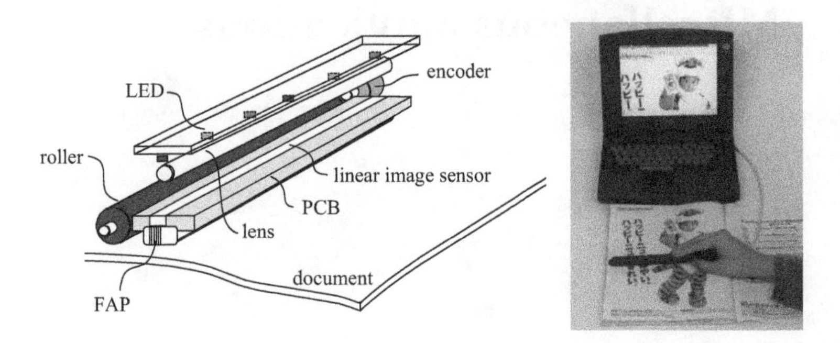

Figure 6.1 Configuration of a compact scanner and its photograph.

(CMOS) driver circuits. It is mounted on a printed circuit board (PCB). A fiber array plate (FAP) is attached to the surface of the image sensor. The other side of the FAP becomes in contact with a document. An encoder is connected to a roller to record the displacement of the image sensor on a document. All components are contained in a 10 mm-diameter housing. By turning on the blue LEDs, a blue image of a linear region of the document is recorded. This is repeated for green and red images. When the encoder detects that the scanner moves a certain predetermined distance, the sequence of image acquisition is repeated for the next line.

Two technologies play important roles in realizing this compact scanner. The first is dedicated optics including the FAP. The second is the image sensors fabricated on glass substrates by LTPS TFT technology. These are described below.

6.1.1 Optical configurations

Inside a facsimile, a document becomes in contact with a module containing a linear image sensor and a light source. The light source illuminates a linear region of the document. The image sensor captures the image of one line on the document at a time. The whole document is digitized by mechanically translating the document. A flat-bed scanner and a copier adopt a similar configuration. A hand-held scanner needs a roller and an encoder to measure the translation distance.

The size of a scanner is mostly determined by its optical configuration. A cross section of a conventional hand-held scanner is schematically drawn in Figure 6.2a. A reduced image of one line of a document is formed on a linear charge-coupled device (CCD) by a lens. To house all components in a small container, its optical path is folded by a mirror. A contact-type configuration shown in Figure 6.2b utilizes a gradient-index lens array [3]. It forms an erect image of the document on a linear image sensor. Hence, it must be as wide as the document. An image sensor fabricated on a glass substrate by thin-film

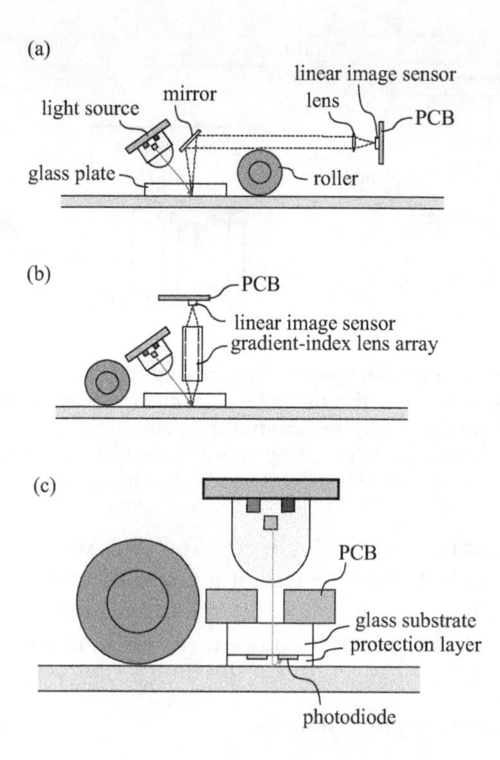

Figure 6.2 Three configurations for a hand-held scanner: (a) an optical reduction system, (b) contact-type scanner, and (c) full-contact scanner.

semiconductor processes is suitable for this configuration. Alternatively, multiple CCDs can be tiled with minimum gaps between them. The distance between the document and the image sensor is called conjugate length. It is typically longer than 10 mm. Full-contact configuration in Figure 6.2c is adopted by a more compact scanner. A document is illuminated through the apertures made in the photodiodes and the PCB. Hence, CCDs cannot be used in this configuration. The thin protection layer in contact with a document provides room for the diffusively reflected light to reach the sensitive region of the photodiode.

Ideally, the protection layer in Figure 6.2c should be closely in contact with the document. However, it is difficult to ensure this condition for a hand-held scanner because it is manually translated against a document. The gap between them allows the light reflected by the document to enter the neighboring photodiodes, leading to degradation of spatial resolution.

To alleviate this problem, the protection layer is replaced by a FAP as shown in Figure 6.3. A FAP is a bundle of optical fibers with light-absorbing materials embedded in their walls. An optical fiber has an acceptance angle beyond which the light cannot propagate inside by repeating total internal

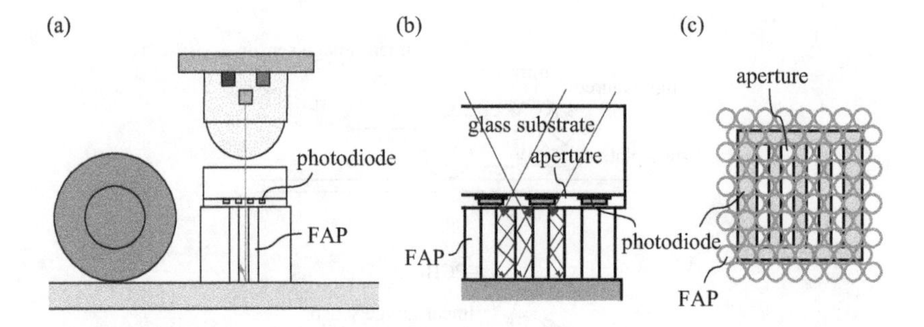

Figure 6.3 An optical configuration for illuminating the document and detecting the light reflected by it through a FAP: (a) cross section of the whole structure, (b) enlarged cross section around the photodiode and the FAP with some trajectories of the illuminating light and the reflected light, and (c) top view of the structure showing the photodiode with apertures and the FAP.

reflection (TIR) at the interface of its core and clad. When the light breaking the TIR condition leaks into the clad, it is absorbed by the wall. Because the FAP absorbs the light propagating obliquely, degradation of spatial resolution is lessened even when the FAP becomes apart from the document.

6.1.2 Image sensors on glass substrates

A photograph of a part of the image sensor is shown in Figure 6.4a. It was observed with a reflected-light microscope. The photograph in Figure 6.4b was observed by transmitted light. It shows the region around the photodiodes after a FAP was attached. Note that there are several optical fibers in an aperture.

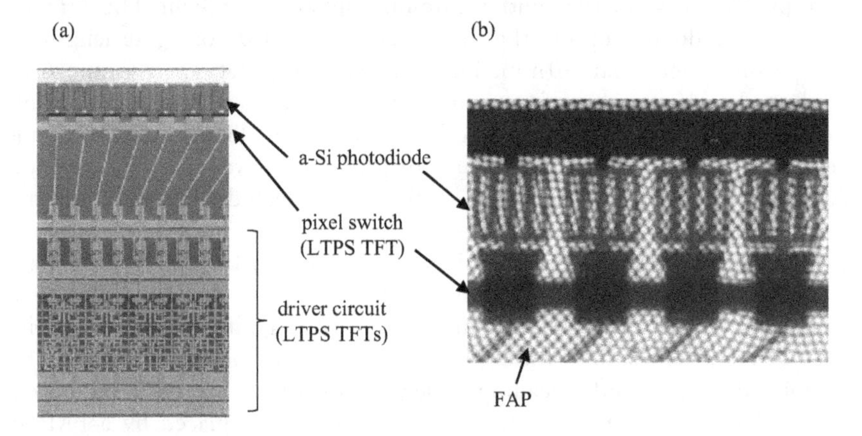

Figure 6.4 Microscope photographs of a part of the linear image sensor fabricated on a glass substrate. (a) Array of a-Si photodiodes and CMOS driver circuit observed by reflected light and (b) the region of the a-Si photodiodes attached to a FAP observed by transmitted light.

6.2 Fingerprint sensors with liveness detection

Although electronic commerce continues to flourish, spoofing with replicated fingers remains a problem for fingerprint authentication [4]. Several methods have been studied to check if a finger is real. For example, vital signs such as skin temperature, electrical resistance, electrocardiogram, pulse oximetry, etc. can be detected when a fingerprint image is acquired. Even laser speckle contrast imaging can be applied for detecting blood flow in a finger [5]. Detection of these signals requires extra hardware.

In this regard, a technique that does not hint at its detection principle is preferred. For example, perspiration starts from pores and diffuses along the ridges when a finger is pressed against a fingerprint sensor. A signal related to this phenomenon can be extracted from two images acquired a few seconds apart. Using a commercial capacitive fingerprint sensor, trade-off between the accuracy of verification and the time interval for image acquisitions was discussed [6].

An optical sensor can also detect vital signs of a finger. In the mid-2000s, it was conceived that color changes in a finger could be used to reject a replicated finger [7]. They are caused by the blood movement in a pressed finger, and they can be detected more quickly than sweating.

6.2.1 Scattered light detection

An optical configuration illustrated in Figure 6.5a detects the light scattered by a finger [8]. An LED is coupled to the edge of a transparent plate. The light emitted by the LED is trapped inside by TIR. The ridges of a finger are in contact with the plate and scatter the light. An erect image of the finger surface is acquired with an array of gradient-index lenses and a two-dimensional image sensor. Note that the light also enters the finger from the ridges. After being diffusively reflected inside the finger, this component is superimposed on the light scattered by the ridges. As shown in Figure 6.5b, they can be called volume and surface components. The latter

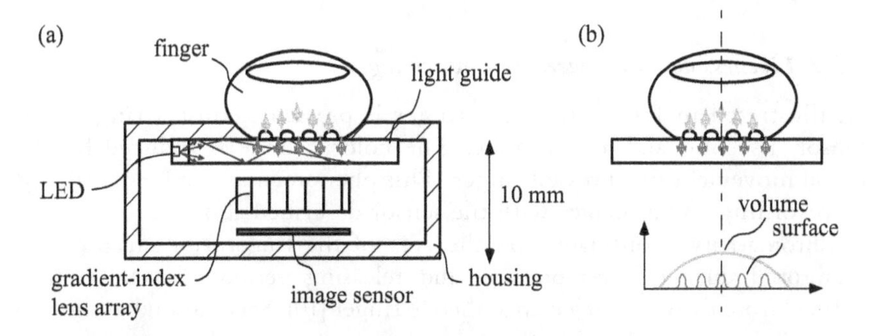

Figure 6.5 Fingerprint sensor based on scattered light detection: (a) configuration and (b) two components in the image.

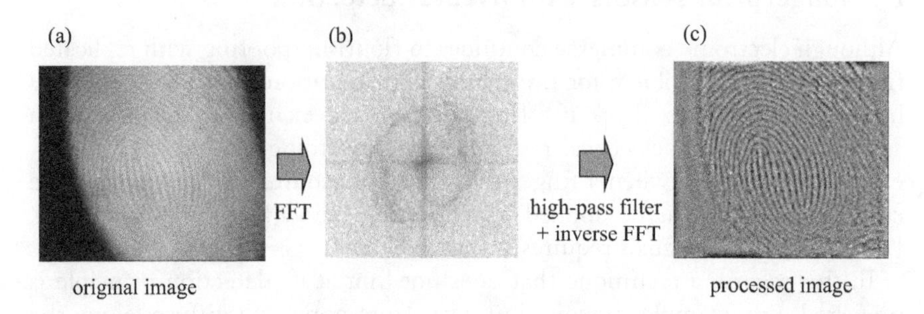

Figure 6.6 An example of enhancing a fingerprint image by eliminating the light scattered inside a finger: (a) an original image, (b) Fourier spectrum of the original image, and (c) the image after filtering and inverse Fourier transform.

carries fingerprint information. Color images are acquired with a white LED and a color image sensor. Alternatively, one can acquire three images with a monochrome sensor by turning on a red, green, and blue LED sequentially.

Although the volume component represents a signal for liveness detection, it is a noise for fingerprint recognition. Because the two components contain different spatial frequency components, a standard image processing technique can separate them. An example of eliminating the low-frequency component with a high-pass filter is shown in Figure 6.6. Fast Fourier transform (FFT) algorithm [9] converts an original image to one in spatial frequency domain. Low-frequency components are set to zero by a high-pass filter. Inverse FFT of the filtered data results in an image with fingerprint information.

The distance between the image sensor and the top of the light guide is equal to the conjugate length of the gradient-index lens (see Section 6.1.1). Such a lens array with a conjugate length of 10 mm was commercially available in the late 1990s. For embedding in a smartphone, another configuration is required. For example, a stack of an OLED display, a collimator, and an image sensor can be adopted (see Section 4.3.1).

6.2.2 Liveness detection based on color changes

As illustrated in Figure 6.7, as a finger is pressed against a fingerprint sensor, its color and area change. The color change is induced by the blood movement in a pressed finger. This phenomenon can be monitored by recording video images with the sensor described above.

Chromaticity coordinates and the area of the fingerprint images were monitored during finger-pressing and releasing periods. Several indices defined from these data rejected a silicone finger [10]. Spectral studies showed that the changes occurred in the red and green wavelength ranges for a live finger. Nine fake fingers made of different materials were tested with a

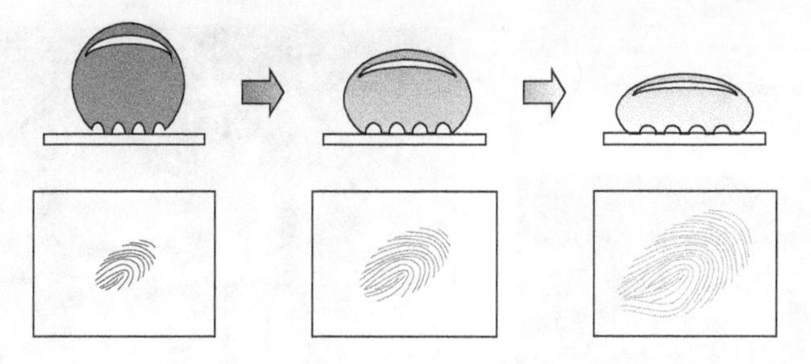

Figure 6.7 Blood movement inside a finger can be monitored by detecting color changes in fingerprint images during an input action.

dual-LED system without a blue LED and all were successfully rejected [11]. Furthermore, hysteresis was observed in the case of living fingers. It indicates that the blood comes back to its original position more slowly. With a proper model, this information might be able to evaluate the stiffness of capillary blood vessels in a finger [12].

6.3 Energy-harvesting displays

Two types of configurations for incorporating photovoltaic technologies into displays are described here. The first example utilizes a luminescent solar concentrator (LSC) as a screen for a projector. The second example simply stacks a semi-transparent display on a solar cell.

6.3.1 Energy-harvesting projector screens

An LSC was proposed in 1976 when solar cells were expensive [13]. A luminescent material is either dispersed in a transparent plate or coated on its surface. Solar cells are attached to its edge surfaces. When sunlight excites the luminescent material, photoluminescence (PL) photons are generated. A large portion of the PL photons are trapped in the plate by TIR at its interface with air. They become concentrated at its edges where solar cells harvest them. An incident area for sunlight can be much larger than the area of a solar cell. Geometrical gain is defined as the ratio of the two areas. The power conversion efficiency of 7.1% and 4.6% are reported for LSCs with a geometrical gain of 2.5 and 10, respectively [14].

Although fabrication cost of solar cells has dropped, interest in LSC technology continues due to its attractive appearance in urban environments: LSCs are semi-transparent and come in various colors. Building-integrated photovoltaic (BIPV) is a concept to generate electric power by covering building roofs, facades, and walls with solar panels in urban environments. An LSC is a promising BIPV system that does not ruin aesthetic values of cities [15]. For example, a "smart window" harvests energy from

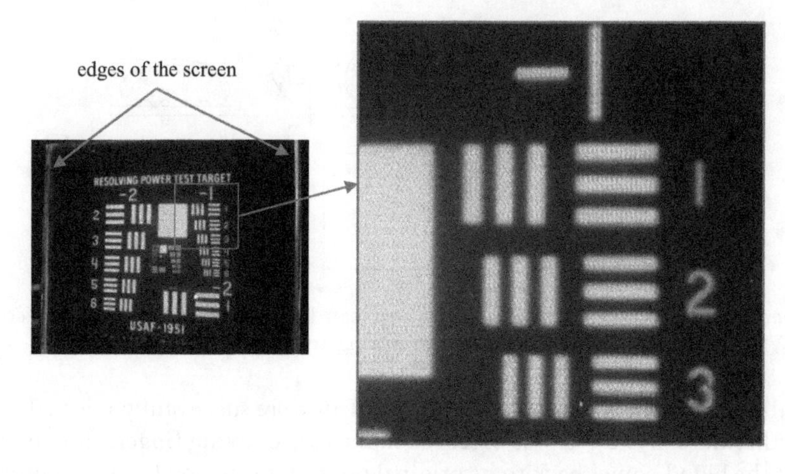

Figure 6.8 An LSC is used as a projector screen for displaying images. In this demonstration, a blue image of a resolution chart is projected on a screen consisting of two acrylic plates sandwiching a thin Coumarin 6 layer. Individual mirrors of the DMD projector can be faintly identified in the magnified photograph.

ambient light and its transmittance is adjusted according to the amount of incident sunlight, room temperature, weather, etc. [16]. This is accomplished by a guest–host system consisting of luminescent dyes and liquid crystal (LC) molecules. When an electric field aligns the LC molecules, the dyes change their orientations with them, resulting in changes in transmittance of ambient light. Light absorption and emission in such a guest–host system can be described by the tilted dipole model [17]. Another smart window concept proposes to combine an LSC and a Venetian window for stand-alone ventilation and temperature control in a building [18].

The PL photons can escape an LSC if the TIR condition is broken at the interface with air. That is why LSCs exhibit various colors. Utilizing these PL photons more actively, one can display images on an LSC by projecting blue images on it [19]. The photograph of a proof-of-concept experiment is shown in Figure 6.8. The screen consists of two acrylic plates sandwiching a thin Coumarin 6 layer. The original LEDs in a digital micromirror device (DMD)-based projector are replaced by a laser diode to project a blue image. The magnified photograph shows that individual mirrors of the DMD are visible. The PL photons escaping from the edges could have been harvested by covering these areas with solar cells. Using three types of luminescent materials, color images can be displayed [20].

6.3.2 Energy-harvesting flat-panel displays

Wristwatches can harvest power from sunlight and scavenge energy from ambient light. In the mid-1990s, Citizen Watch Co. Ltd. commercialized a solar-powered wristwatch and named its technology as Eco-Drive [21].

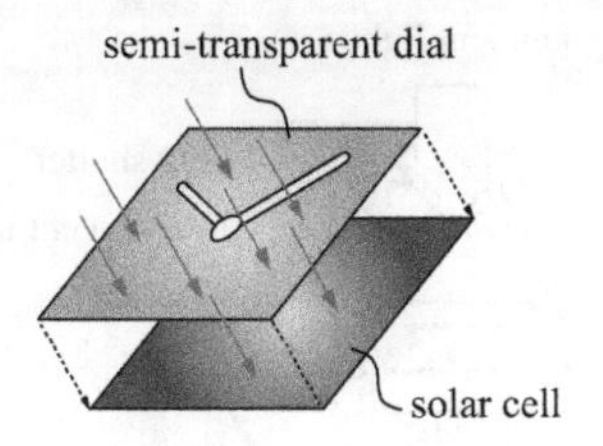

Figure 6.9 A solar cell is placed beneath a semi-transparent dial of a wristwatch.

As illustrated in Figure 6.9, a solar cell is placed beneath a semi-transparent dial. An energy-storage cell stores the converted electric power. Thus, the wristwatch can continue to operate with intermittent exposure to light.

Now that a wristwatch displays various images, a mobile display might be able to power itself by harvesting energy from ambient light. One might call it a display-integrated photovoltaic system. In fact, an energy-harvesting OLED display was reported in 2008 [22]. An OLED becomes almost transparent when its top and bottom electrodes are made of transparent materials. As shown in Figure 6.10, a solar cell is placed beneath such an OLED display. A circular polarized is usually attached to the surface of an OLED display to suppress reflection of ambient light (see Section 5.2.3). It is not needed by this configuration because the solar cell absorbs the incident light. In addition to the ambient light passing through the OLED display, half of the light emitted by the OLEDs propagates toward the solar cell. It recovers a part of the energy of this downward photon flux.

Reflection of ambient light at the surface of an OLED display cannot be eliminated completely. Hence, its ambient contrast ratio degrades in bright environments. In this regard, a luminous-reflective display (LRD) utilizes ambient light. Because its luminance is proportional to illuminance at its surface, ambient contrast ratio remains constant irrespective of illuminance (see Section 5.4). Adopting the Eco-Drive concept, a slight modification enables an LRD to harvest energy from ambient light [23]. As shown in Figure 6.11, the

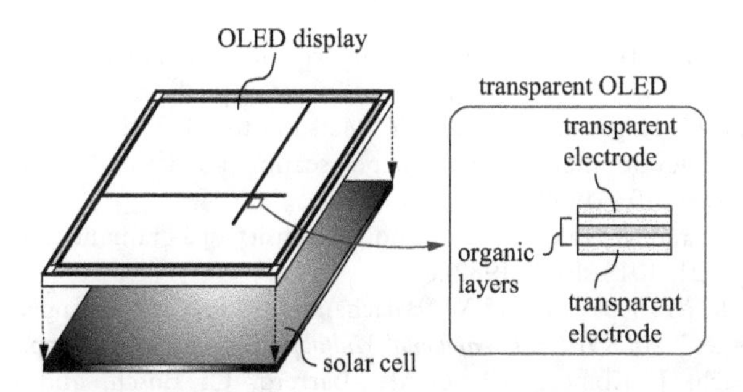

Figure 6.10 A solar cell placed beneath a transparent OLED display harvests energy from ambient light passing through them.

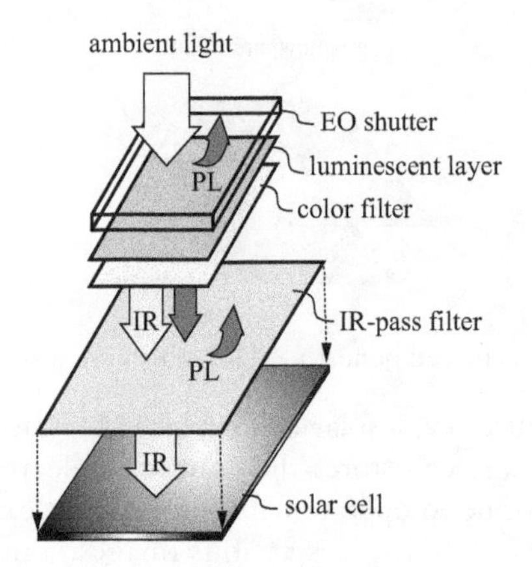

Figure 6.11 Configuration of a sub-pixel in an energy-harvesting LRD.

reflector at each sub-pixel in an LRD is replaced by an infrared (IR)-pass filter, and a solar cell is placed beneath it.

Operation principle is as follows. When the electro-optic (EO) shutter is set to on-state, ambient light passes through it. Visible light within a specific wavelength range excites the luminescent layer for each sub-pixel while IR photons are transmitted unabsorbed. Hence, the solar cell can harvest the IR photons in the ambient light. In the experiment, luminescent layers stacked on an infrared pass filter were excited by monochromatic light from an edge-lit front-light unit at 405 nm. A one-dimensional model reproduced the measured spectral fluxes reasonably well [24].

A transmissive monochrome LC panel can be used as the EO shutter array. A color LC panel is transparent in the near-IR wavelength range [25]. This fact proves that an LC layer, a polarizer film, and color filters are all transparent in this wavelength range. Therefore, IR photons can reach the solar cell when the EO shutter is set to on-state. There are a lot of near-IR photons in solar spectra.

References

1. M. Gao, X. Hu, B. Cao, and D. Li, "Fingerprint sensors in mobile devices," *2014 9th IEEE Conference on Industrial Electronics and Applications*, 2014, pp. 1437–1440.
2. I. Fujieda, H. Haga, F. Okumura, Y. Matsumoto, H. Kohashi, H. Matsuo, and S. Miura, "Development of a pen-shaped scanner and its applications," *Opt. Eng.* 38(12), 2093–2103 (1999).
3. M. Toyama and M. Takami, "Luminous intensity of a gradient-index lens array," *Appl. Opt.* 21, 1013–1016 (1982).
4. S. Memon, M. Sepasian, and W. Balachandran, "Review of finger print sensing technologies," *2008 IEEE International Multitopic Conference*, 2008, pp. 226–231.
5. P. Keilbach, J. Kolberg, M. Gomez-Barrero, C. Busch, and H. Langweg, "Fingerprint presentation attack detection using laser speckle contrast imaging,"

2018 International Conference of the Biometrics Special Interest Group (BIOSIG), 2018, pp. 1–6.

6. R. Derakhshani, S. A. C. Schuckers, L. A. Hornak, and L. O'Gorman, "Determination of vitality from a non-invasive biomedical measurement for use in fingerprint scanners," *Pattern Recognit.* 36(2), 383–396 (2003).
7. I. Fujieda, E. Matsuyama, and M. Kurita, "Signatures of live fingers extracted from a series of fingerprint images," *Proc. SPIE* 5677, 177–185 (2005).
8. I. Fujieda and H. Haga, "Fingerprint input based on scattered light detection," *Appl. Opt.* 36(35), 9152–9156 (1997).
9. R. C. Gonzalez and P. Wintz, *Digital Image Processing*, Addison-Wesley, 1977, p. 78.
10. K. Tai, M. Kurita, and I. Fujieda, "Recognition of living fingers with a sensor based on scattered-light detection," *Appl. Opt.* 45(3), 419–424 (2006).
11. K. Tai, E. Matsuyama, M. Kurita, and I. Fujieda, "Dual-LED imaging for finger liveliness detection and its evaluation with replicas," *Appl. Opt.* 45(24), 6263–6269 (2006).
12. A. Hori and I. Fujieda, "Studies on blood movement during a fingerprint input action," *Int. J. Optomechatron.* 2(4), 390–400 (2008).
13. W. H. Weber and J. Lambe, "Luminescent greenhouse collector for solar radiation," *Appl. Opt.* 15(10), 2299–2300 (1976).
14. L. H. Slooff, E. E. Bende, A. R. Burgers, T. Budel, M. Pravettoni, R. P. Kenny, E. D. Dunlop, and A. Büchtemann, "A luminescent solar concentrator with 7.1% power conversion efficiency," *Phys. Status Solidi RRL* 2, 257–259 (2008).
15. F. Meinardi, F. Bruni, and S. Brovelli, "Luminescent solar concentrators for building-integrated photovoltaics," *Nat. Rev. Mater.* 2, 17072 (2017).
16. M. G. Debije, "Solar energy collectors with tunable transmission," *Adv. Funct. Mater.* 20(9), 1498–1502 (2010).
17. I. Fujieda, D. Suzuki, and T. Masuda, "Tilted dipole model for bias-dependent photoluminescence pattern," *J. Appl. Phys.* 116, 224507 (2014).
18. N. Aste, M. Buzzetti, C. Del Pero, R. Fusco, F. Leonforte, and D. Testa, "Triggering a large scale luminescent solar concentrators market: The smart window project," *J. Clean.* 219, 35–45 (2019).
19. I. Fujieda, S. Itaya, M. Ohta, Y. Hirai, and T. Kohmoto, "Energy-harvesting laser phosphor display and its design considerations," *J. Photon. Energy* 7, 028001 (2017).
20. K. Yunoki, R. Matsumura, T. Kohmoto, M. Ohta, Y. Tsutsumi, and I. Fujieda, "Cross talk and optical efficiency of an energy-harvesting color projector utilizing ceramic phosphors," *Appl. Opt.* 58, 9896–9903 (2019).
21. A. Harris, "Watching the clock," *Eng. Technol.* 4(9), 60–61 (2009).
22. C-J. Yang, T-Y. Cho, C-L. Lin, and C-C. Wu, "Energy-recycling high-contrast organic light-emitting devices," *J. Soc. Inf. Display* 16(6), 691–694 (2008).
23. S. Matsuda, H. Nishimura, Y. Mizuno, and I. Fujieda, "Luminance, color gamut, and energy-harvesting characteristics of luminescent layers placed above a solar cell," *Opt. Express* 29, 36784–36795 (2021).
24. T. Anekawa, M. Shigeta, and I. Fujieda, "Front lighting for an energy-harvesting luminous-reflective display and its design considerations," *Opt. Eng.* 61(9), 095102 (2022).
25. T. Large, N. Emerton, S. Lim, and C. Nuesmeyer, "Far field imaging through liquid crystal displays for biometrics," *SID Symp. Dig. Tech. Pap.* 52, 210–213 (2021).

7 Appendix

A1 Standard theories for semiconductor devices

A1.1 Abrupt junction approximation

When a p-type semiconductor is in contact with an n-type semiconductor, carriers are depleted at the junction. As shown in Figure A1.1a, donor atoms and acceptor atoms in this region are positively and negatively charged, respectively. Suppose that they are completely ionized and that their distributions along the depth direction x change abruptly. Then, the charge

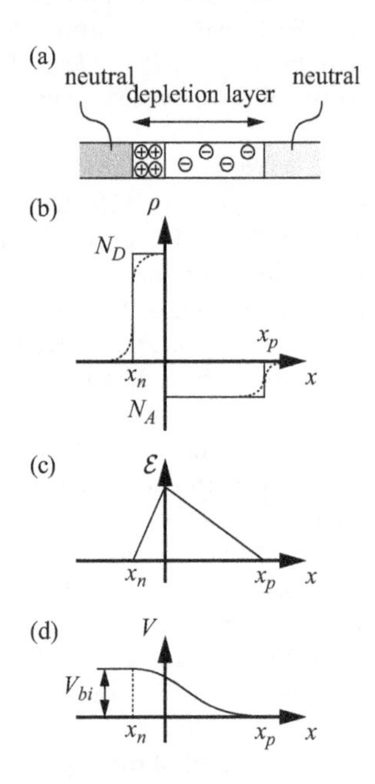

Figure A1.1 Abrupt junction approximation for a p–n junction. (a) Mobile charges are depleted near the junction. (b) Fixed charge density is assumed to change abruptly at each interface of adjacent regions. (c) Electric field and (d) potential distributions.

density distribution ρ can be represented by the solid lines in Figure A1.1b where doping concentrations are denoted as N_D and N_A.

Electric field $\mathcal{E}$, space charge density ρ, and permittivity ε are related by Maxwell's equation as follows.

$$\frac{d\mathcal{E}}{dx} = \frac{\rho}{\varepsilon} \tag{A1.1}$$

From the definition of a potential V,

$$-\frac{dV}{dx} = \mathcal{E} \tag{A1.2}$$

Eliminating $\mathcal{E}$ from these equations results in Poisson's equation below.

$$\frac{d^2V}{dx^2} = -\frac{\rho}{\varepsilon} \tag{A1.3}$$

Example A1.1

Assume the following charge density distribution and boundary condition.

$$\rho = \begin{cases} qN_D & x_n \le x < 0 \\ -qN_A & 0 < x \le x_p \end{cases}, \quad \begin{cases} V(-x_n) = V_{\mathrm{bi}} \\ V(x_p) = 0 \end{cases}$$

Note that x_n is a negative number.

Find the field and the electric field and potential distributions.

Express the depletion layer width $W = -x_n + x_p$ in terms of built-in potential V_{bi} and dopant concentration N_D and N_A.

Solution

Integrating $\frac{d\mathcal{E}}{dx} = \frac{\rho}{\varepsilon}$, the electric field in each region is given by,

$$\mathcal{E} = \begin{cases} \frac{qN_D}{\varepsilon} x + C_1 & x_n \le x < 0 \\ -\frac{qN_A}{\varepsilon} x + C_2 & 0 < x \le x_p \end{cases}$$

Constant of integration is determined by setting the electric field to zero at $x = x_n, x_p$. Hence,

$$\mathcal{E} = \begin{cases} \frac{qN_D}{\varepsilon}(x - x_n) & x_n \le x < 0 \\ -\frac{qN_A}{\varepsilon}(x - x_p) & 0 < x \le x_p \end{cases}$$

Electric field must be continuous at $x = 0$.

$$\therefore -N_D x_n = N_A x_p$$

Integrating $-\dfrac{dV}{dx} = \mathcal{E}$, electric potential in each region is given by,

$$V = \begin{cases} -\dfrac{qN_D}{2\varepsilon}\left(x - x_n\right)^2 + C_3 & x_n \le x < 0 \\ \dfrac{qN_A}{2\varepsilon}\left(x - x_p\right)^2 + C_4 & 0 < x \le x_p \end{cases}$$

Using the boundary condition $V_{bi} = C_3$, $0 = C_4$, the final expression is as follows.

$$V = \begin{cases} -\dfrac{qN_D}{2\varepsilon}\left(x - x_n\right)^2 + V_{bi} & x_n \le x < 0 \\ \dfrac{qN_A}{2\varepsilon}\left(x - x_p\right)^2 & 0 < x \le x_p \end{cases}$$

The electric potential must be continuous at $x = 0$. Setting x to zero in the equations above results in the following expression for the built-in field.

$$V_{bi} = \frac{q}{2\varepsilon}\left(N_D x_n^2 + N_A x_p^2\right)$$

Using $-N_D x_n = N_A x_p$, this is rewritten as,

$$V_{bi} = \frac{q}{2\varepsilon}\left[N_D\left(-\frac{N_A}{N_D}x_p\right)^2 + N_A x_p^2\right] = \frac{q}{2\varepsilon}N_A\left(1 + \frac{N_A}{N_D}\right)x_p^2$$

Using V_{bi}, depletion width is expressed as follows.

$$W^2 = \left(x_p - x_n\right)^2 = \left(1 + \frac{N_A}{N_D}\right)^2 x_p^2 = \left(1 + \frac{N_A}{N_D}\right)\frac{V_{bi}}{\frac{q}{2\varepsilon}N_A}$$

$$= \frac{2\varepsilon}{q}\left(\frac{1}{N_A} + \frac{1}{N_D}\right)V_{bi}$$

$$\therefore W = \sqrt{\frac{2\varepsilon}{q}\left(\frac{1}{N_A} + \frac{1}{N_D}\right)V_{bi}}$$

A1.2 Gradual channel approximation

Let us consider an n-channel MOSFET illustrated in Figure A1.2 and formulate drain current I_d. The coordinate system is as defined. Suppose that the source is grounded. When the gate bias V_g exceeds the threshold voltage V_{th}, a thin sheet of mobile charges (electrons in this case) appears in the region $0 \le x \le L$, $0 \le z \le W$, and $y \approx 0$. This region is called a channel. Its length and width are denoted as L and W, respectively. The depletion layer extends below the channel as well as at the two p–n junctions.

When a small positive bias is applied to the drain, the potential at the channel gradually increases from 0 to V_d along the x axis over the distance L. The potential distribution along the y axis is much steeper because the dimension along the y axis is much smaller than L. The potential distribution along the z axis is uniform. Thus, only the potential distribution along the x axis is of interest. In such a case, the analysis reduces to a one-dimensional problem and this simplification is called gradual channel approximation.

The gate and the channel sandwiching the gate insulator constitute a capacitor. Denoting this capacitance per unit area as C_{ox} and the channel voltage at x as $V(x)$, the charges per unit area at x is expressed as $Q_{ch}(x) = C_{ox}\{V_g - V_{th} - V(x)\}$. Note that the charges at x per unit length along the x axis is $WQ_{ch}(x)$. Because the electric field at x is expressed as $-\frac{dV}{dx}$, the velocity of the channel charges $WQ_{ch}(x)$ is equal to $\mu_{fe}\left(-\frac{dV}{dx}\right)$, where μ_{fe} is called field-effect mobility. Current is simply the product of the charges and their velocity. Therefore, the drain current is expressed as follows.

$$I_d = -W\mu_{fe}C_{ox}\left(V_g - V_{th} - V\right)\frac{dV}{dx} \tag{A1.4}$$

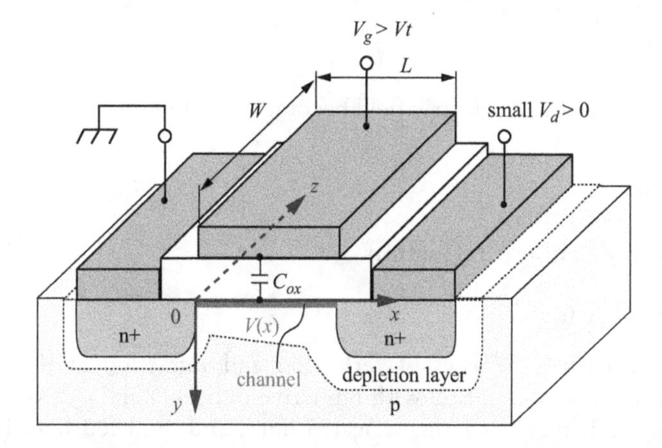

Figure A1.2 Geometry of an n-channel MOSFET and the coordinate system for formulating its current–voltage characteristics.

Suppose that the channel charge $Q_{ch}(x)$ exists in the region $0 \leq x \leq L$. Integrating the equation above in this region,

$$\int_0^L I_d dx = -W\mu_{fe}C_{ox}\int_0^{V_d}\left(V_g - V_{th} - V\right)dV \tag{A1.5}$$

Noting that I_d is constant, performing this integration leads to the following expression.

$$I_d = -\mu_{fe}C_{ox}\frac{W}{L}\left\{\left(V_g - V_{th}\right)V_d - \frac{V_d^2}{2}\right\} \tag{A1.6}$$

The negative sign means that the current flows from the drain to the source. Note that this equation is valid only if the channel charge $Q_{ch}(x)$ extends from the source to the drain. Setting $Q_{ch}(x) \geq 0$, this condition is expressed as $V_d \leq V_g - V_{th}$.

At $V_d = V_g - V_{th}$, the channel charge $Q_{ch}(x)$ vanishes at $x = L$. This condition is called pinch-off.

The channel voltage $V(x)$ is derived by setting the upper limit for integration to x and $V(x)$. Solving the resulting quadratic equation, it is expressed as,

$$V(x) = \left(V_g - V_{th}\right)\left(1 - \sqrt{1 - \frac{x}{L}}\right) \tag{A1.7}$$

The electric field is given by,

$$\mathcal{E}(x) = -\frac{dV}{dx} = \left(V_g - V_{th}\right)\frac{1}{2L\sqrt{1 - \frac{x}{L}}} \tag{A1.8}$$

As x approaches L, the electric field becomes very high.

A2 Optics

A2.1 Refraction and reflection

A2.1.1 Snell's law

Let us summarize the laws of reflection and refraction at the interface between two isotropic media with refractive indices n_1 and n_2. As illustrated in Figure A2.1, incident light, reflected light, and refracted light lie in the same plane, which is called the plane of incidence. The law of reflection states that the angle of reflection is equal to the angle of incident θ_1. If the interface

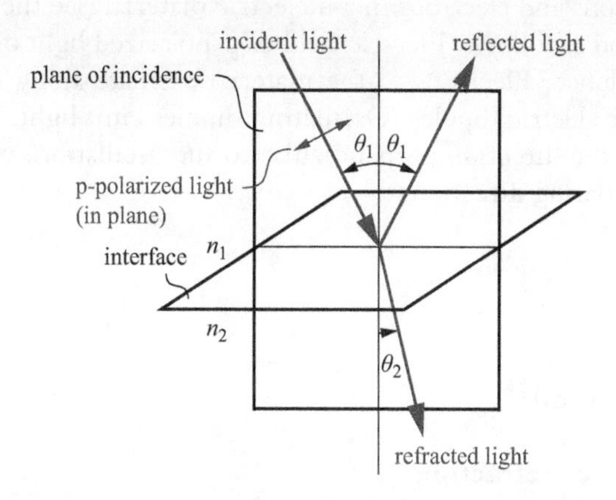

Figure A2.1 Definitions of the plane of incidence and s- and p-polarized light.

is flat, θ_1 has only one value. This is called specular reflection. If the interface is rough, the light is reflected to various directions. This is called diffuse reflection. The angle of refraction θ_2 is related to the angle of incident θ_1 by Snell's law as follows.

$$n_2 \sin\theta_2 = n_1 \sin\theta_1 \tag{A2.1}$$

In general, incident light is unpolarized. It can be decomposed into two components perpendicular to each other. One whose electric field oscillates in the plane of incidence is called p-polarized. The other component is called s-polarized.

A2.1.2 Fresnel equations

The probability of reflection at the boundary is given by Fresnel equations. They are given by the equations below for the s- and p-polarized components, respectively.

$$R_S = \frac{\sin^2(\theta_1 - \theta_2)}{\sin^2(\theta_1 + \theta_2)} \tag{A2.2}$$

$$R_P = \frac{\tan^2(\theta_1 - \theta_2)}{\tan^2(\theta_1 + \theta_2)} \tag{A2.3}$$

The probability R_P becomes zero at a particular angle given by the equation below. This is called Brewster's angle. In the context of energy transfer between photons and electrons in a dielectric material (see the next section), it is understood as follows. Electric field of p-polarized light oscillates in the plane of incidence. Electrons in the material oscillate along this field, and they represent electric dipoles. Oscillating dipoles emit light. Its intensity is maximum in the direction perpendicular to the oscillation, while it is zero along the oscillating direction.

$$\theta_B = \tan^{-1}\frac{n_2}{n_1} \tag{A2.4}$$

A2.2 Polarization

A2.2.1 Index of refraction

Light propagation in a material is explained in the context of energy transfer between photons and electrons. An incident photon exerts Coulomb force on electrons in a material. They oscillate synchronously to the vibrating electric field of the incident photon. Electrons are displaced from their equilibrium positions, and the atoms become polarized. Oscillating dipoles emit photons. Light propagates in a material by repeating this process. The speed of light is smaller in a material than in vacuum. Refractive index is defined as this ratio. Because atoms in a crystal are arranged with fixed interatomic distances in three-dimensional space, a crystal has three axes in general. Hence, a crystal exhibits different refractive indices depending on the direction of the oscillating electric field. This optical property of a material is called birefringence. On the other hand, optical properties of amorphous materials are isotropic because they have no axes.

A2.2.2 Birefringence

Let us consider a crystal with only one axis. Its optical property is characterized by two refractive indices. The Fresnel ellipsoid illustrated in Figure A2.2 is a virtual object in three-dimensional space that tells us the refractive index of a uniaxial crystal. The long and short intercepts are refractive indices for extraordinary rays and ordinary rays, respectively. For example, linearly polarized light labeled "e-ray" propagates along the x axis. We regard that its electric field intersects the ellipsoid at $z = n_e$. This is interpreted as the refractive index that this material exhibits for this polarized light. The ray labeled as "o-ray" sees a material with refractive index n_o. For rays propagating along the z axis, the material exhibits n_o, irrespective of their polarization states. For the case of oblique incidence with polar angle θ, the refractive index is calculated by,

$$n = \frac{n_e n_o}{\sqrt{n_e^2 \cos^2\theta + n_o^2 \sin^2\theta}} \tag{A2.5}$$

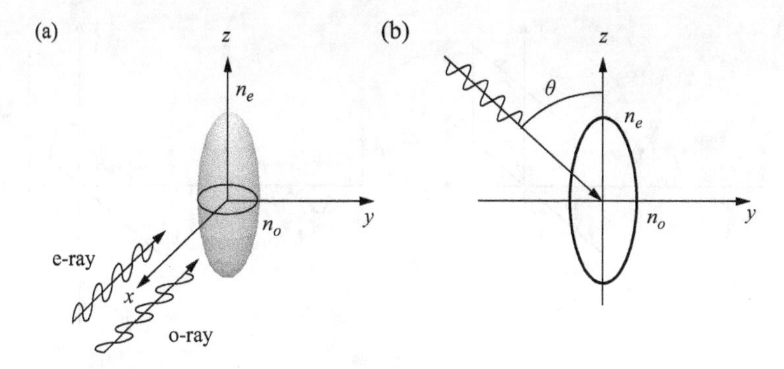

Figure A2.2 Fresnel ellipsoid. (a) Rays enter along the x axis. (b) Cross section of Fresnel ellipsoid with an obliquely incident light.

A2.2.3 Jones vector

Suppose that a plane wave is propagating along the z axis. For an observer receiving this wave, the two components of its electric field are expressed as $A_x e^{i(kz-\omega t+\phi_x)}$ and $A_y e^{i(kz-\omega t+\phi_y)}$. When we are interested in phase only, we can drop the factor $e^{i(kz-\omega t)}$. The Jones vector for this wave is defined as $\begin{bmatrix} A_x e^{i\phi_x} \\ A_y e^{i\phi_y} \end{bmatrix}$. It is customary to express the x component by a real value. This vector is equivalent to $\begin{bmatrix} A_x \\ A_y e^{i(\phi_y-\phi_x)} \end{bmatrix}$. Two special cases for $A_x = A_y$ are illustrated in Figure A2.3. In Figure A2.3a, the two components are in phase, i.e., $\phi_x = \phi_y$. The tip of the electric field oscillates along the line $y = x$. The normalized Jones vector is $\frac{1}{\sqrt{2}}\begin{bmatrix} 1 \\ 1 \end{bmatrix}$. This is an example of linearly polarized light. If the vertical component is delayed by 90°, the tip rotates counterclockwise for the observer receiving this wave as illustrated in Figure A2.3b. This is a left-handed circular polarized light. In this case, $e^{i(\phi_y-\phi_x)} = e^{i\pi/2} = i$. Hence, its normalized Jones vector is $\frac{1}{\sqrt{2}}\begin{bmatrix} 1 \\ i \end{bmatrix}$. Table A2.1 summarizes normalized Jones vector in case of $A_x = A_y$.

A2.2.4 Jones matrix

Jones matrices are operators that change the polarization states. This is expressed as,

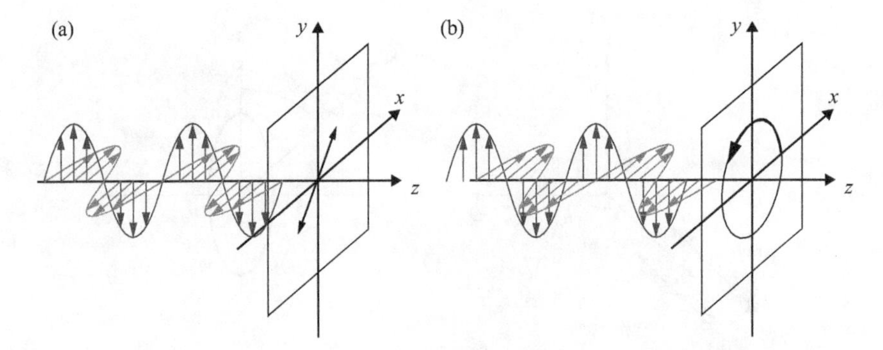

Figure A2.3 Electric field vector consists of two orthogonal components. (a) When there is no delay between them, the tip of the electric field vector moves along a line. (b) When the delay is 90°, its trajectory is an ellipsoid in general. If the two components have the same amplitude, it is a circle.

Table A2.1 Normalized Jones vector in case of $A_x = A_y$

Linearly polarized			*Circularly polarized*	
Along x	*Along y*	*In the 45° direction*	*Right-handed*	*Left-handed*
$\begin{bmatrix} 1 \\ 0 \end{bmatrix}$	$\begin{bmatrix} 0 \\ 1 \end{bmatrix}$	$\frac{1}{\sqrt{2}}\begin{bmatrix} 1 \\ 1 \end{bmatrix}$	$\frac{1}{\sqrt{2}}\begin{bmatrix} 1 \\ -i \end{bmatrix}$	$\frac{1}{\sqrt{2}}\begin{bmatrix} 1 \\ i \end{bmatrix}$

$$\vec{U}' = L\vec{U} = \begin{bmatrix} l_{11} & l_{12} \\ l_{21} & l_{22} \end{bmatrix} \vec{U} \tag{A2.6}$$

Jones matrices for polarizers and waveplates are summarized in Table A2.2.

A3 Color science

The International Commission on Illumination (Commission Internationale de l'éclairage, CIE) is an organization that standardizes vision, color,

Table A2.2 Jones matrix

Polarizer		*Quarter-waveplate*		*Half-waveplate*
Trans. axis: x	*Trans. axis: y*	*Slow axis: x*	*Slow axis: y*	
$\begin{bmatrix} 1 & 0 \\ 0 & 0 \end{bmatrix}$	$\begin{bmatrix} 0 & 0 \\ 0 & 1 \end{bmatrix}$	$\begin{bmatrix} 1 & 0 \\ 0 & -i \end{bmatrix}$	$\begin{bmatrix} 1 & 0 \\ 0 & i \end{bmatrix}$	$\begin{bmatrix} 1 & 0 \\ 0 & -1 \end{bmatrix}$

measurements, etc. It defines various functions to relate the physical properties of light and how we perceive color.

A3.1 Photometry

Photopic and scotopic luminous efficiency functions represent the spectral sensitivity of our eyes under bright and dim lighting conditions, respectively. They are plotted as a function of wavelength λ in Figure A3.1.

Photometric quantities $\phi_V(\lambda)$ are calculated from radiometric quantities $\phi_e(\lambda)$ and the luminous efficiency function $V(\lambda)$ by the following equation. Table A3.1 summarizes various corresponding quantities.

$$\phi_V = K_m \int_{380}^{780} \phi_e(\lambda) \cdot V(\lambda)\, d\lambda, \quad K_m = 683\,\text{lm/W} \tag{A3.1}$$

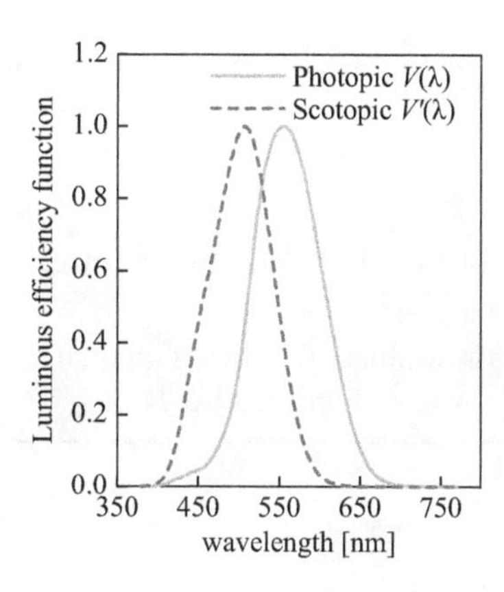

Figure A3.1 Luminous efficiency functions used for calculating various photometric quantities.

Table A3.1 Radiometric and photometric quantities

Radiometric quantities	*Unit*	*Photometric quantities*	*Unit*
Radiant flux	W	Luminous flux	lm
Radiant intensity	W/sr	Luminous intensity	cd
Radiance	$W/sr \cdot m^2$	Luminance	cd/m^2, nit
Irradiance	W/m^2	Illuminance	lx
Radiant exitance	W/m^2	Luminous exitance	lm/m^2, rlx

Example A3.1

A monochromatic point source emits a radiant flux of 1.00 W in all directions uniformly. Its wavelength is 555 nm. An observer is in a bright environment.

1. Calculate the luminous flux of this light source. Hint: replace ϕ_e by a δ function because the source is monochromatic.
2. Calculate its radiant intensity and luminous intensity?
3. An illuminance meter is placed 1.00 m away from this point source. What is its output?
4. What is its output when the distance is increased to 2.00 m?

Solution

1. $\phi_V = K_m \int_{380}^{780} \delta(555)\cdot V(\lambda)\,d\lambda = K_m V(555) = 683\ \text{lm}$
2. The light spreads over the solid angle 4π sr. Hence, its radiant intensity is $\dfrac{1.00\ \text{W}}{4\pi\ \text{sr}} = 0.07957\cdots = 0.0796\ \text{W/sr}$. Its luminous intensity is $\dfrac{683\,\text{lm}}{4\pi\,\text{sr}} = 54.4\ \text{cd}$.
3. A luminous flux of 683 lm lands on the area of $4\pi\ \text{m}^2$. Hence, the illuminance is $54.4\ \text{lm/m}^2 = 54.4\ \text{lx}$.
4. When the distance is doubled, the area is quadrupled. Hence, illuminance is reduced by a factor of 4. Hence, 13.6 lx.

A3.2 Color spaces

A3.2.1 Color matching functions

Color is perceived by three types of photosensors in our retinas. Based on the spectral sensitivity of these sensors, CIE defined three functions $\bar{x}(\lambda)$, $\bar{y}(\lambda)$, and $\bar{z}(\lambda)$ in 1931, such that $\bar{y}(\lambda)$ coincides with the photopic luminous efficiency function. These are called CIE 1931 color matching functions and are plotted in Figure A3.2.

A3.2.2 Tristimulus values

Denoting the spectrum of a light source as $S(\lambda)$ and the spectral reflectance of an object as $\rho(\lambda)$, the spectrum of the light reflected by the object is equal to $S(\lambda)\cdot\rho(\lambda)$. When this light enters our eyes, our photosensors produce three signals called tristimulus values. They are expressed as follows.

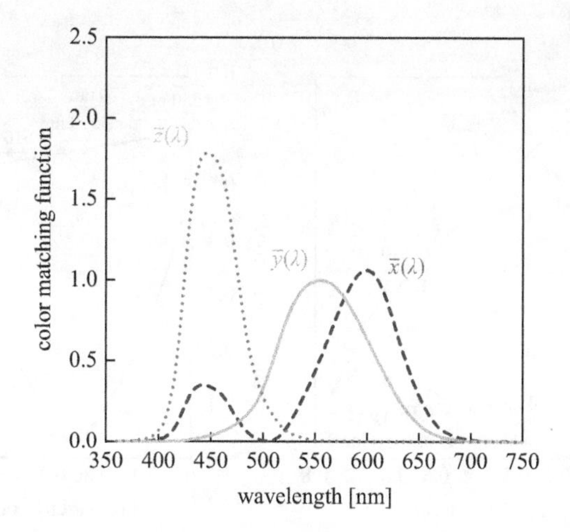

Figure A3.2 CIE 1931 color matching functions.

$$\begin{cases} X = K\int_{\lambda} S(\lambda)\cdot\rho(\lambda)\cdot\bar{x}(\lambda)d\lambda \\ Y = K\int_{\lambda} S(\lambda)\cdot\rho(\lambda)\cdot\bar{y}(\lambda)d\lambda \\ Z = K\int_{\lambda} S(\lambda)\cdot\rho(\lambda)\cdot\bar{z}(\lambda)d\lambda \end{cases} \tag{A3.2}$$

where K is a proportional constant. There are two definitions for K as expressed below. The first definition sets the maximum value of Y to 100. The second definition relates the tristimulus value Y to photometric quantities.

$$K = \frac{100}{\int_{\lambda} S(\lambda)\cdot\bar{y}(\lambda)d\lambda} \quad \text{or} \quad K = 683\ [\text{lm/W}] \tag{A3.3}$$

Because color is represented by a set of three values, it can be regarded as a vector. The three-dimensional space based on the color matching functions above is called the CIE 1931 XYZ color space.

A3.2.3 Chromaticity diagrams

A vector has two attributes: magnitude and direction. In the same manner, a color vector has two attributes: the strength of the stimulus and chromaticity. Hence, the chromaticity coordinates (x, y) are defined below for specifying chromaticity.

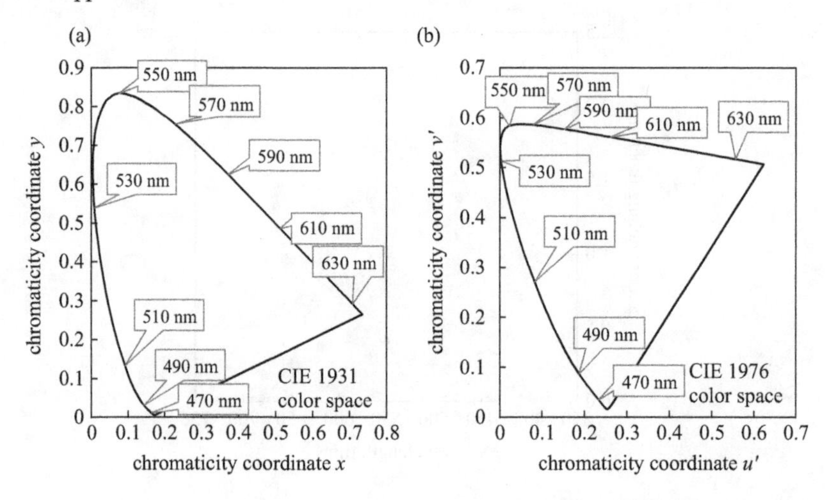

Figure A3.3 The chromaticity diagrams: (a) CIE 1931 xy chromaticity diagram and (b) CIE 1976 $u'v'$ chromaticity diagram.

$$x = \frac{X}{X+Y+Z}, \quad y = \frac{Y}{X+Y+Z}, \quad z = \frac{Z}{X+Y+Z} \tag{A3.4}$$

Because $z = 1 - x - y$, x and y are used for specifying chromaticity. As shown in Figure A3.3a, all colors we perceive reside either inside or on the horseshoe-shaped region. The curved edge is the trajectory for monochromatic light. It is called spectral locus. This plot is known as the CIE xy chromaticity diagram. Another diagram defined by CIE in 1976 is shown in Figure A3.3b.

It is convenient to define the difference between two colors as the distance between them in a chromaticity diagram. A trace of the threshold distances where we can distinguish two colors resembles an eclipse. It is called McAdam eclipse. The problem with the CIE 1931 xy chromaticity diagram is that MacAdam ellipses depend on (x, y). For example, the ellipses for green colors are much larger than those for red and blue colors. To lessen this problem, CIE defined Uniform Color Space (UCS) in 1960. Currently, its modified version called CIE 1976 L*u*v* (abbreviated as CIELUV) color space is gaining popularity. Its chromaticity coordinates (u', v') are defined as follows. They are plotted in Figure A3.4b.

$$u' = \frac{4x}{-2x+12y+3}, \quad v' = \frac{9y}{-2x+12y+3} \tag{A3.5}$$

A3.2.4 Color gamut

Additive color mixing is regarded as the addition of vectors in the CIE 1931 XYZ color space as illustrated in Figure A3.4. A color vector intercepts the plane $X+Y+Z=1$ at (x, y, z). Suppose that $(x_1, y_1, 0)$ and $(x_2, y_2, 0)$ are

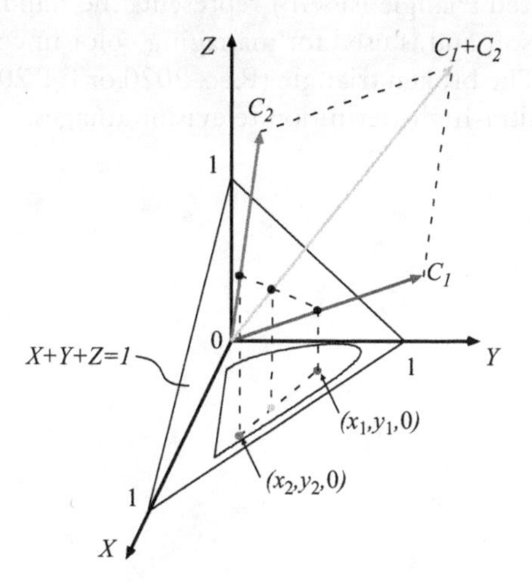

Figure A3.4 Additive color mixing is regarded as the addition of vectors in a three-dimensional space.

the chromaticity coordinates of the vectors C_1 and C_2. The chromaticity coordinate of the vector $C_1 + C_2$ is an internally dividing point of the line connecting these two points on the *xy* plane. Therefore, one can create all points on this line by varying the magnitudes of the constituent color vectors.

By adding three primary colors, one can create all colors inside and along the triangle whose apexes are the chromaticity coordinates of the constituent colors. This triangle is called a color gamut. For managing colors among different modalities such as displays and printers, it is convenient to define a standard color gamut. Some examples are shown in Figure A3.5. The solid triangle is the standard set by the National Television System Committee

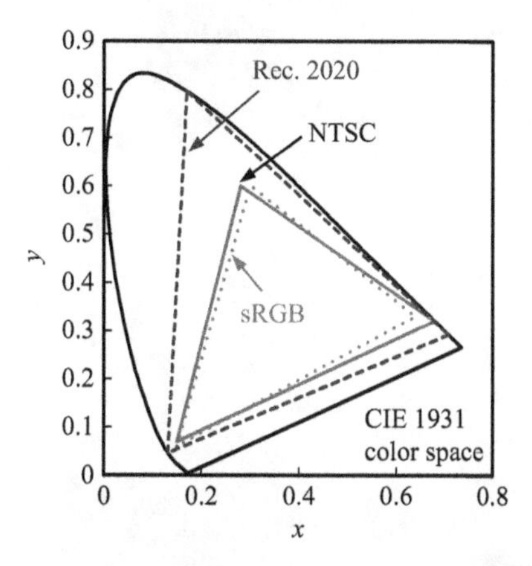

Figure A3.5 Standard color gamut is compared in the CIE1931 chromaticity diagram.

(NTSC). The dotted triangle (sRGB) represents the standard created jointly by HP and Microsoft and is used for managing color in computers, printers, and the Internet. The broken triangle (Rec. 2020 or BT 2020) is the standard for broadcasting ultra-high-definition television images.

Index